T0283418

WORLD BANK TECHNICAL PAPER NUMBER 168

Solar Pumping

An Introduction and Update on the Technology, Performance, Costs, and Economics

Roy Barlow, Bernard McNelis, and Anthony Derrick

Intermediate Technology Publications and
The World Bank, Washington, D.C.

Copyright © 1993
The International Bank for Reconstruction
and Development/THE WORLD BANK
1818 H Street, N.W.
Washington, D.C. 20433, U.S.A.

Published in the United Kingdom, 1993, by
Intermediate Technology Publications, Ltd.,
103/105 Southampton Row, London, WC1B 4HH, U.K.

Technical Papers are published to communicate the results of the Bank's work to the development community with the least possible delay. The typescript of this paper therefore has not been prepared in accordance with the procedures appropriate to formal printed texts, and the World Bank accepts no responsibility for errors.

The findings, interpretations, and conclusions expressed in this paper are entirely those of the author(s) and should not be attributed in any manner to the World Bank, to its affiliated organizations, or to members of its Board of Executive Directors or the countries they represent. The World Bank does not guarantee the accuracy of the data included in this publication and accepts no responsibility whatsoever for any consequence of their use. Any maps that accompany the text have been prepared solely for the convenience of readers; the designations and presentation of material in them do not imply the expression of any opinion whatsoever on the part of the World Bank, its affiliates, or its Board or member countries concerning the legal status of any country, territory, city, or area or of the authorities thereof or concerning the delimitation of its boundaries or its national affiliation.

The material in this publication is copyrighted. Requests for permission to reproduce portions of it should be sent to the Office of the Publisher at the address shown in the copyright notice above. The World Bank encourages dissemination of its work and will normally give permission promptly and, when the reproduction is for noncommercial purposes, without asking a fee. Permission to copy portions for classroom use is granted through the Copyright Clearance Center, 27 Congress Street, Salem, Massachusetts 01970, U.S.A.

The complete backlist of publications from the World Bank is shown in the annual *Index of Publications*, which contains an alphabetical title list (with full ordering information) and indexes of subjects, authors, and countries and regions. The latest edition is available free of charge from the Distribution Unit, Office of the Publisher, Department F, The World Bank, 1818 H Street, N.W., Washington, D.C. 20433, U.S.A., or from Publications, The World Bank, 66, avenue d'Iéna, 75116 Paris, France.

ISSN: 0253-7494
ISBN: (U.S. edition) 0-8213-2101-3
ISBN: (U.K. edition) 1 85339 179 4

At IT Power, Ltd. Roy Barlow is Project Engineer, Bernard McNelis is Managing Director, and Anthony Derrick is Director.

Library of Congress Cataloging-in-Publication Data

Barlow, Roy, 1965–
    Solar pumping : an introduction and update on the technology,
performance, costs, and economics / Roy Barlow, Bernard McNelis, and
Anthony Derrick.
       p.   cm. — (World Bank technical paper ; no. 168)
    Includes bibliographical references.
    ISBN 0-8213-2101-3
    1. Solar pumps. 2. Water-supply engineering—Developing
countries. 3. Irrigation engineering—Developing countries.
I. McNelis, Bernard. II. Derrick, Anthony, 1954– . III. Series.
TJ912.B37   1992
628.1'44—dc20   92-21847
                                                CIP

FOREWORD

Over the past fifteen years the solar pump has evolved to become a technology which can make a significant contribution to water supply in the rural areas of developing countries. There are around 5,000 solar pumps in use around the world, and the 1990s should see a major increase in numbers. This Guide hopes to assist with this process.

The work which has lead to this guide covers more than a decade. In the mid-1970s many initiatives to use solar energy were launched. Solar powered pumping was seen as an important possibility. In 1978 the United Nations Development Programme began funding a project which was initiated by the World Bank, for the 'testing and demonstration of small scale solar powered pumping systems'. The programme was designed to assemble reliable technical and economic data from which to form a considered view of the viability of solar pumping systems. The project team examined the state of existing technology, carried out laboratory tests and field trials, and analysed the results over a period of four years. The project culminated in the production of a 'Handbook on Solar Water Pumping', published by the World Bank (Washington) in 1984. This was updated and published as a book by I T Publications (London) in 1985.

Since completion of this work, and largely as a result of the UNDP/World Bank solar pumping project, small solar pumps have been developed to be simple to install and reliable in use. In many instances solar pumps can provide water at a cost lower than other water pumping technologies. The World Bank has continued to follow developments in solar pumping, and has sponsored the preparation of this Guidebook.

The Guidebook has been written to provide both a simple introduction to potential users and purchasers of solar pumps, and to review the current state of the art in technology, costs and economics. The principal author is Roy Barlow with co-authors Bernard McNelis and Anthony Derrick of I T Power Ltd.

Richard Stern
Manager
Energy Sector Management Assistance Program

TABLE OF CONTENTS

TABLES

1. INTRODUCTION

1.1. Purpose of this book

Water is one of the primary resources necessary to support life, as has been so tragically demonstrated in recent years by several droughts in the Sahel region of Africa. Even in regions where the rainfall does not fluctuate so severely, access to a clean and reliable water supply can make a vital difference to the health and quality of life of a rural community. In many of these areas water exists below the ground, and throughout the developing world the most widespread way of raising it to the surface is still by handpump or with the assistance of animals. The principal mechanised power source is the diesel engine, but this is often beyond the means or technical capability of small communities.

Although the number of units in the field is still small, solar photovoltaic powered pumping systems offer many advantages over the more traditional technologies. Because there are few moving parts, maintenance is reduced to a minimum, and reliability is very high. Also, because the time of greatest water demand usually coincides with the maximum daily solar energy, the available pumping power is well matched to the demand.

Solar pumping was first introduced into the field in the late nineteen-seventies, and since then manufacturers have refined their products to give considerable increases in performance and reliability. The steady fall in prices of solar photovoltaic (PV) panels means that solar pumping is becoming economic for an increasingly wide range of applications.

This guide is written for the potential user to give a simple background to PV pumping technology and to help to identify the situations in which solar pumping should be considered.

The early chapters act as a guide to those unfamiliar with PV pumping, illustrating typical applications and reviewing current technology. Following this are sections dealing with the range of currently available equipment and examining experience in the field. The final four sections cover the practical aspects of choosing a pumping system. This includes site evaluation and system sizing, a simple methodology for an economic assessment, and advice on procurement, installation and maintenance.

The appendices contain various data and information referred to in the text. Two of these are of particular note : Appendix G, which contains quick reference data for wind, diesel and hand pumping scenarios; and appendix I which reviews the current and future economics of PV pumping in general in comparison with other pumping alternatives.

1.2. The first decade of PV pumping

A significant part of the early development of solar technology was concerned with water pumping. A solar steam engine pumped water at the Paris Exposition in 1878. The development and commercialisation of a viable solar-thermodynamic pump was pioneered by the French company

Sofretes following early work in Senegal. In the 1970's many of these pumps were installed around the world, particularly in the African Sahel and Mexico.

Development of photovoltaic pumps was also pioneered in France. The first systems used a Pompes Guinard motor-pump comprising a surface-mounted permanent-magnet DC motor driving a submersed centrifugal pump via a vertical shaft. The motor was connected directly to the PV array. Several demonstration units were installed in the 1970's.

In 1978 a team at the World Bank presented a compelling case for a programme to apply small PV pumps for irrigation ("micro-irrigation") on a huge scale. A goal of 10 million units installed by the year 2000 was presented as appropriate, as having a significant impact on world food production yet representing only 10 percent of potential farmer users. Economic analysis suggested that small PV pumps would be economic with array costs of around $5/Wp (1978 dollars) which was the US Department of Energy projected price for 1981.

As a result of the work mentioned above, as well as other interests, the United Nations Development Programme (UNDP) provided funding for the Global Solar Pumping Project (GLO/78/004), to be executed by the World Bank. In July 1979 consultants were appointed by the World Bank to evaluate, test and demonstrate commercially available small-scale solar powered irrigation pumping systems. At the time the technical feasibility of solar powered pumping had been demonstrated in a limited way, but the technology was clearly immature and expensive. The purpose of the project was to advise the UNDP and the World Bank on the way in which solar pumps should be developed and applied so as to provide an appropriate method for irrigation under the conditions that prevail on small farms in developing countries. The capacity required amounted to flows in the range one to five litres per second and static heads up to 7 m. Pumps of this capacity have outputs in the range 150 to 500 W and can irrigate areas of between 0.5 and 1.0 hectares, depending on the crop and efficiency of water distribution. The target cost of water delivered was $0.05 per cubic metre (1979 prices). When the project was defined both solar-thermodynamic and photovoltaic technologies were considered to be of equal merit and potential.

Following the review a state-of-the-art report was released in 1979. This was followed by an international call for offers, together with 250 questionnaires sent to potential suppliers. This resulted in only 13 suppliers being able to realistically supply equipment. A total of twelve pumps were tested in the field, four in each of Mali, Sudan and the Philippines. One system was solar thermodynamic the remainder were photovoltaic.

Field testing was conducted in 1980, and PV modules, pumps and motors were also tested as components in the UK and USA. Three systems performed better than their rated value, two systems were within 10% of rated performance, and five systems performed significantly below rating. Two systems (including the solar thermodynamic pump) failed to operate.

The detailed results of this work were published in 1981 and concluded that there was indeed considerable potential for the use of solar pumps for irrigation, but that none of the products then available on the market were yet suitable for widespread use. Extensive recommendations were made on improving performance, reliability and cost. The project also reported that meeting the target irrigation water cost of $0.05/m^3$ would not be easy but that a significantly higher cost would be acceptable for village water supply.

The UNDP and World Bank decided to continue the work to the point where really widespread demonstration (Phase II) could be contemplated. Important new activities included :

- Assessment of prospective countries for participation in Phase II (countries short-listed were Bangladesh, Brazil, Egypt, Kenya, Mexico, Pakistan, Sri Lanka and Thailand)

- Procurement of improved commercial systems and sub-systems for testing in order to qualify them for use in Phase II.

Based on the experience gained in Phase I, exacting specifications were developed to define the performance of improved systems, and tenders were invited.

Of 64 systems tendered 12 were purchased and laboratory tested in the UK in 1982/83. The final report was released in 1983 and this concluded performance had improved but that there was still the need and scope for improvements in performance and reliability. From the economic studies it was concluded that PV pumps were broadly competitive with the primary alternatives and could be justified in sunny regions where diesel costs are high, wind speeds low and a steady year round demand for water exists. It was shown that village water supply would in general become economic before irrigation.

The final output of the work was the Handbook on Solar Water Pumping which was published by the World Bank in 1984. The World Bank commissioned an update on PV pumping in 1986 and this Guide results from further work in 1989/1990. Given the lead of the UNDP/World Bank project, PV pumps have, over the past decade, evolved to be simple and reliable. A total of more than 10,000 have been installed to date, of which 30-40% are in developing countries.

2. WHEN TO USE SOLAR PUMPS

2.1. Energy requirements for water pumping

The starting point for any assessment of solar water pumping is to determine the energy requirement for your pumping needs. The greater the energy demand, the more PV modules will be needed, and so the greater the cost. Some understanding of how energy requirements are calculated is therefore important.

The energy used in lifting a certain amount of water (hydraulic energy) is directly proportional to both the volume of water lifted (V in m³), and the 'head' (the height it is lifted through, h in metres). The most convenient unit of energy is the kilowatt hour (kWh). This is simply the energy equivalent to 1000 W for 1 hour. The hydraulic energy, E, in kWh can be calculated from

$$E = Vh / 367$$

For example, to lift a volume of 60 m³ through a height of 10 m requires 60 X 10 / 367 = 1.64 kWh.

However, not all the energy produced by the PV modules will be available as hydraulic energy. Each time energy is converted from one form to another (i.e., solar to electrical, electrical to mechanical, mechanical to hydraulic), there is an associated loss, and so the input energy must be far greater than the hydraulic energy that is required. This is true of any pumping system and is illustrated by the energy flow diagram in figure 2.1. As pumping equipment is usually purchased from the supplier as a complete system, the user will not normally need to consider efficiencies of individual components. The performance is more likely to be expressed as the volume per day delivered at a given head with a given daily solar energy.

In addition to energy conversion losses, a proportion of the pumped water may be lost in the process of delivering it to its point of use. This must be taken account of when calculating the water requirement, as it will have a direct effect on the energy required for pumping, and thus the cost of the system. To minimise the cost and maximise the performance it is therefore important for all the components in the system to be as efficient as possible.

The term 'head' is a common one when dealing with solar pumping, and it is important to have a clear understanding of its meaning. The head can be thought of as consisting of two parts: The static head is simply the height over which the water is pumped, and can be easily measured, as shown in the schematic of a typical pumping system in figure 2.2. The dynamic head is due to the increased pressure caused by friction of the water through the pipework. It is expressed as an equivalent height of water, and can be calculated from the flow rate, pipe size and pipe materials. Smaller pipes and higher flow rates produce a greater dynamic head. The total head, used to calculate the hydraulic energy requirement, is the sum of the static and dynamic heads. The total head is proportional to the hydraulic energy requirement, with the result that it is cheaper to pump through lower heads.

Figure 2.1: Schematic of energy flows in a PV pumping system

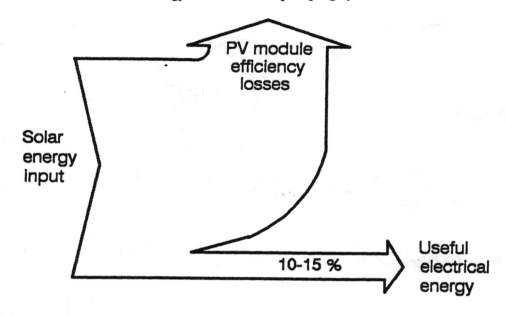

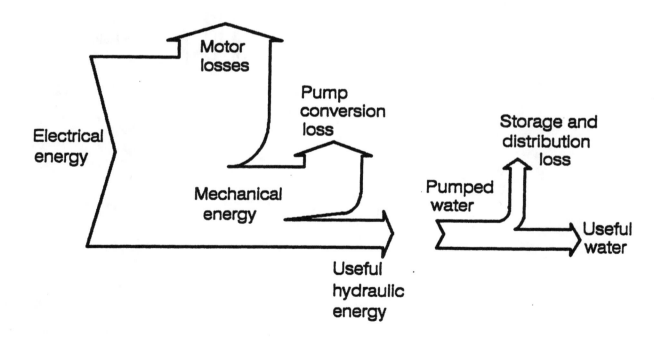

Figure 2.2: Schematic of a typical rural pumping system

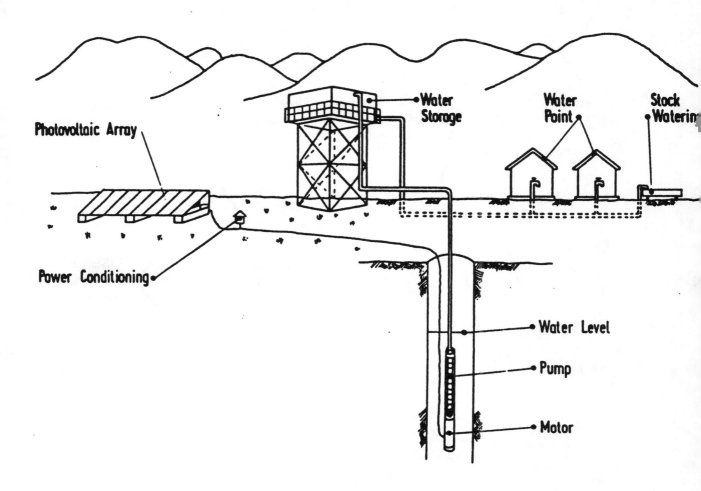

6

Another commonly used quantity is the volume-head product, Vh. This is simply the volume per day (in m^3) multiplied by the total head (in m). The units of the volume-head product are m^4/day, and pump performance is sometimes expressed this way as it is directly proportional to the daily energy requirement. The unit m^4 can also be expressed as $m^3.m$ = Tonne.metres.

2.2. Solar energy

The strength of the sun is measured in Watts per m^2, and is called the irradiance. Just outside the Earth's atmosphere this has a value of 1353 W/m^2, but as the light travels through the atmosphere it is attenuated due to absorption by dust, etc., and becomes weaker with increasing atmospheric path. The total power falling on a horizontal surface (the Global Irradiance) reaches a maximum of about 1000 W/m^2 at sea level. The can be thought of as two components, the radiation in the direct beam from the sun, and diffuse radiation from the sky (i.e., radiation that has been scattered by clouds or the atmosphere).

The global irradiance varies throughout the course of the day due to changes in sun angle and thus path length through the atmosphere. For the same reason there are variations with season and latitude, with changing length of day an additional factor. The total solar energy received in a day (known as the solar insolation or solar irradiation) can vary from 0.5 kWh/m^2 in a northern winter, to 6 kWh/m^2 in the tropics. It is fortunate that the semi-arid areas which have the greatest water demand are also those with some of the highest insolation figures. Maps showing the global distribution of solar insolation for each season can be found in appendix A.

On a clear day, diffuse radiation may account for about 15% of the total, whereas it would be close to 100% on an overcast day. Even on completely cloudy days there may well be enough diffuse radiation to continue pumping, although direct sunlight will obviously give better performance. Thus compared to many more conventional power sources, the performance of solar pumping systems are quite variable as they are affected by the prevailing weather conditions. However, in the long term the daily insolation is a well defined parameter for a particular area, and is much less site specific than some competing appropriate technologies such as wind pumping.

2.3. Typical water pumping applications

Because of the relatively high cost of PV modules, solar pumping is most economical for small power demand applications, say less than 1000 W. This is well matched with the type of pumping loads that are the most appropriate in the developing world. The major demand for water supply falls into two fairly distinct categories:

(i)   Village water supply
(ii) Irrigation

Both of the above two are often combined with livestock watering where there is sufficient capacity. Solar pumping for irrigation is very rarely competitive in purely economic terms, although many operational installations exist around the world. Therefore, while irrigation pumping will be covered, most attention will be given to village supply situations. Although not rigidly defined, a village in this context is assumed to be a rural community, remote from the electricity grid.

Solar pumping is better suited to village water supply, both because of the smaller quantities of water involved (and thus lower power demand) and the comparatively high value of domestic water compared to that for irrigation. Water demand tends to be roughly constant throughout the year, but it is very desirable to incorporate a covered storage tank in the system (to guard against periods of poor supply or pump breakdown). In practical terms storage for more than a day or two is rarely feasible, and a backup handpump may be cheaper than a larger tank in some cases (e.g., on open wells).

A typical example would be a supply for a village of 500 people. The World Health Organisation (WHO) have defined a desirable per capita water consumption for the developing world of 40 litres per day (although the survival level is far below this). This 40 litres is intended to cover drinking, washing, cooking and sanitation. Thus the whole village would require a total of 20 m^3 per day (assuming, of course, that the well could yield this quantity). To avoid contamination in surface water, drinking water is usually taken from boreholes or wells, and so a realistic head might be 20 m. Therefore using the formula in section 2.1 shows that an hydraulic energy of 1.1 kWh per day is necessary. If a pumping day is around 8 hours, this gives an average power demand of 140 W. This is well within the region in which PV pumping is cost effective. Large animals, such as horses or cattle, require a similar per capita supply to humans. So a system supplying a herd of 500 cattle would be of similar size to that for a typical village. When considering the design of a system the logistics of distribution (e.g., taps, stand-pipes) must also be considered to make most effective use of the water, and the amount of the users' time taken in collecting water.

Water for irrigation is characterised by a large variation from month to month in the amount of water required. To get the maximum benefit from irrigation the crop may need its maximum water supply during quite a short period. Growth rate may be sensitive to any deviations from this pattern, and so demand may peak at 100 m^3/day/hectare in some months and drop to zero in others. Thus for most of the year the pump is greatly oversized for the requirements. In irrigation applications it is necessary to pump more water than is actually used, to overcome inefficiencies in the water distribution system and field application methods. Because of their larger scale these are inherently more 'leaky' than the pipework systems for village water supply. Generally it is not economic to lift water for irrigation through anything but very low heads or from surface water. This is because increasing the lift increases the cost, and the cost of supplying water for irrigation should not be more than the value of the additional crop that can be grown. In practice, PV irrigation can usually only be competitive for micro-irrigation schemes for vegetable gardens and plots of less than about 1 ha. The well yield is just as likely to be a limiting factor on the irrigated area. Because of the more even month-by-month supply, and the higher value of the water, solar pumps for village water supplies can be competitive at far higher lifts than those for irrigation.

2.4. Physical and economic factors

Since a solar pump consists essentially of an electrical source powering a motor/pump unit, it is technically possible to use a solar pump in all applications where an electric pump can be used. The assessment of viability therefore centres on the question of costs. Although the decision-making process is dealt with in more detail in later chapters, we will give a brief overview here of the methodology for assessing solar pump viability.

When deciding on a power source for your water pumping system there are several basic factors that are needed to decide how the costs of various alternative methods of pumping compare.

Location dependent factors:

- Minimum average monthly windspeed in m/s (u)
- Minimum average monthly solar radiation in $kWh/m^2/day$ (H)

Water demand factors:

- Daily average (crop, village or livestock) water requirement by month in cubic metres per day (V).
- Water table depth in metres (d). This should be monthly if there is seasonal variation.
- The total lift (h), which is the sum of water table depth, tank height, and the expected lowering of the well water level due to pumping (well drawdown).
- It is also useful to calculate (by the month) the average volume-head product in m^4/day, which is h x V.

(If these are not immediately known, methods for obtaining them are described or referenced in later chapters).

As a general approximation it can be shown that solar pumping systems for irrigation begin to become cost effective compared to diesel pumps in situations where the peak daily water requirements (in terms of the volume-head product) are less than about 250 m^4 (e.g., 50 m^3/day through a head of 5 m), and where monthly average solar radiation is greater than 4 $kWh/m^2/day$. For relatively windy locations, windpumps can provide a cheap, reliable water supply. However, windspeed data are often unreliable and the cost of water from a windpump is very sensitive to mean monthly windspeed. The wind regime is also extremely site dependent on scales down to a few hundred metres. Obviously hills and open areas are preferable to sheltered woods or valleys. In situations where the minimum monthly average windspeed is greater than 3 m/s a windpump would probably be the more economic option.

For village supply applications the criteria are less strict. Here, solar pumping systems become cost competitive compared to diesel pumps where the average daily water requirements are less than a volume-head product of up to 800 m⁴ (e.g., 40 m³/day through a 20 m head, and where the monthly average solar irradiation is greater than 2.8 kWh/m²/day. Windpumps are generally cost competitive at locations with minimum monthly average windspeeds greater than 2.5 m/s.

In some circumstances it may be advisable to chose diesel for the short-term, but reconsider solar pumping as a long-term option. Over the last decade, as PV technology and manufacturing methods have improved, there has been a steady fall in the real cost of PV modules, and thus also of water pumping systems. If this trend continues, as is expected, then at some point in the future solar pumping may become economic for situations in which diesel is presently marginally cheaper. A major factor affecting the viability of alternative pumping sources is the local cost of diesel fuel. Although the international market price is reasonably constant, most governments place some sort of tax on diesel, and so the price to the user may vary from about 0.20 up to 1.50 $/litre.

The reader should also be aware of the use of hybrid systems, which make use of a combination of two of the above systems. For instance, in a solar-diesel hybrid the diesel pump would only be needed when demand exceeded what the solar pump could produce or in periods when sunlight levels were too low for normal pumping. These systems tend to be expensive, as one half of the system is lying idle for much of the time. Hybrid systems involving solar power are rarely used in rural scenarios, and their analysis is beyond the scope of this book.

There are, of course, other important factors not captured in a simple cost comparison, and these should be taken into account when comparing pumping systems and making choices. For example, when considering a diesel, the reliability of fuel supplies and the availability of spare parts should be considered. A poor fuel supply may mean that a pump is inoperative when it is needed most, which may have serious consequences. Thus in certain circumstances (e.g., remote, inaccessible locations) the greater reliability of a certain pumping system may offset its higher cost. Another factor is the familiarity of the user community with the technology used for pumping. The people using the equipment must have the necessary technical ability and confidence to perform any routine maintenance tasks. Although PV technology will be unfamiliar to most rural communities, almost zero maintenance is required, and experience has shown that acceptance is generally very good.

The initial criteria for selecting a system can be expressed in their simplest form in the box below, and are also represented in the simple decision flowcharts in figures 2.3. and 2.4. Obviously, these are only a rough guide, and the more detailed assessment methods of later chapters should be used before accepting or discounting a particular pumping method. A general economic comparison between PV, wind, diesel and hand for the village water supply case can be found in appendix I.

Figure 2.4: Decision chart for irrigation

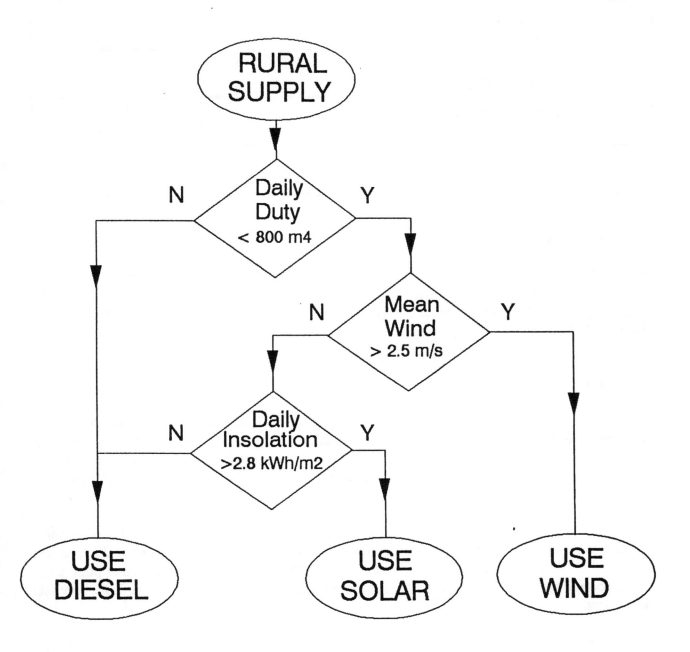

Table 2.4: Decision table for irrigation

		Village	Irrigation	
1.	If max daily duty less than consider wind or solar	800	250	m^4
2.	If min mean windspeed more than consider using wind	2.5	4.0	m/s
3.	If daily insolation more than consider using solar	2.8	3.0	kWh/m^2
4.	If solar or wind not suitable consider using diesel			

2.5. Case Studies

Because PV is still a fairly young technology it is not easy to define a typical pumping installation or programme. However, to give the reader some feel for the way PV pumping is used in the developing world, the following pages briefly review the status of PV pumping in four representative areas: Mali, Morocco, Brazil and India.

2.5.1. Mali

Mali provides a good example of solar pumping experience, and many of the results quoted throughout section 5 of this document have been taken from the Mali PV pumping programme. Data was collected between May and October 1989, and the purpose of the study was to learn from the experiences gained by Mali in installation, finance, operation and maintenance of PV pumping systems.

Due to the large seasonal variation in surface water availability, Mali has many large pumping projects, with a total of around 15,000 boreholes and wells. Most of these are handpumps and around 2000 are diesel or kerosene powered. Comparatively few (around 200) are solar pumps, but where several boreholes are required for hand pumps it is often cheaper to use a single borehole solar pump.

Mali Aqua Viva (MAV), an NGO founded by Pere Vesperen, pioneered the introduction of solar pumps into Mali in 1977, and their success has led to other organisations (USAID, UNDP, GTZ) adopting PV as a power source for pumping. A national organisation, the Cellule d'Entretien des Equipments Solaires (CEES) was created in 1987 to co-ordinate PV pumping under the supervision of the Direction Nationale de l'Hydraulique et de l'Energie (DNHE), which is funded by the French government.

Figure 2.5: Village PV pump in Tioribougou, Mali

Eighty percent of PV pumping in Mali is from boreholes and about 15% from surface water. At the time of writing there were 157 PV pumping system in Mali, and a full list appears in the working report 'Learning from Success' (see bibliography). The number of sites established each year shows an increasing trend, and total installed array capacity is estimated at 220 kWp. Figure 2.5. shows a PV pumping system in Tioribougou in south-western Mali.

The most popular configuration is that of a submersible motor/pumpset, accounting for 43% of installations. Until recently the second most popular was the surface motor/submerged pump, which still features in 25% of present installations. The remainder are either surface, floating or positive displacement devices. Heads are typically in the range 30-40 metres. Lower heads are also common, but there are only 15 sites with head greater than 40 m. The power ratings range from 160 Wp to 12960 Wp for the largest installation, with the mean around 1500 Wp. Outputs are in general found to be consistent with the manufacturers predictions.

The chief suppliers of modules are France Photon, Photowatt and Arco Solar, accounting between them for nearly 70% of the total installed power. The main subsystems suppliers are Guinard and Grundfos with 25 % each, closely followed by Total with 18%. However, a breakdown of subsystems suppliers since 1988 shows that Guinard have dropped out of the market entirely. This is due to problems with their surface motor units (Alta-X) which were replaced after 2 - 5 years due to their poor design. However, the introduction of AC submerged pumps in 1980 led to greater durability, and this is the preferred configuration for new boreholes.

The high investment costs for PV pumps mean that outside donors will have to be involved with the financing of village PV pumping schemes for the foreseeable future. However, the villagers accept the principle that they should contribute in some way, and for MAV installations the beneficiaries contribution to the capital costs now stands at 20%. Villagers must also pay for maintenance and repair. The GTZ installations require a contribution of $3000 out of $36,000 for multi-use pumps, and $5400 out of $20,000 for floating irrigation pumps. Borehole costs are generally financed by the donor institutions.

It has been found that in general families are willing to support a down-payment of $100, and annual payments of $150 per family for a clean and reliable water supply. For comparison the cost of a heifer in Mali was $200 when these figures were compiled.

2.5.2. Morocco

More than 100 solar pumps have been installed by the Ministry of the Interior, and there are probably up to 100 other systems installed privately. The Ministry of the Interior in its plan for 1988-1992 intends to equip 2,000 sites with solar pumps.

Data on installation has been collected by a project conducted by the Centre de Developpement des Energies Renouvelables (CDER) in Marrakech, with funding from USAID. Most installations are rated at around 1kWp, and pump through heads of 15-50m. An important component of the USAID funded work has been the installation of 14 systems in the Tata region of the Anti Atlas.

These use Grundfos and McDonald pumps and some of the systems use tracking PV arrays. Performance data is being collected. An example of a pumping system in Morocco is shown in figure 2.6.

In general Morocco is viewed as a very important potential market.

2.5.3. Brazil

Brazil along with India and China is one of the few developing countries with photovoltaic manufacture and systems integration in the country. The organisation manufacturing PV modules and systems is Heliodinamica of San Paulo which entered the PV market in 1981. By 1989 the company had achieved sales of 600 kWp per annum with revenues exceeding $US 2.6m. However the market for PV pumps has not developed in Brazil with only approximately 53 systems installed by September 1990. The smallest being a 175 Wp and the largest 3.78 kWp (for the North East of Brazil). Arrays ratings of less than 500 Wp account for 50% of the systems installed. Figure 2.7. shows a 420 Wp Brazilian pumping system that produces water for about 40 families.

Imports to the country have been few in number with most PV pumping systems being imported by non-government organisations on which no statistical data exists.

The large livestock herds of the country would suggest a need for rural water pumping and there are several windpump manufacturers in Brazil such as the "Kenya" wind pump manufacturer near Belo Horizonte.

The principal reasons suggested for the poor take up of PV pumping systems have been lack of disposable income during recent economic problems, and no central government orders.

There are plans to install some 8 German manufactured PV pumps under a German-Brazilian cooperation programme, although this is being fiercely contested by the indigenous PV pumping company, Heliodinamica, who manufacture surface water and borehole PV pumping systems and whom have achieved exports to the UK, Germany and India.

In September 1990 the price of a 175 Wp surface water PV pump was $4,555 and for a 1050 Wp borehole pump $15,790.

15

Figure 2.6: (Top) PV pump installation in Morocco
Figure 2.7: (Bottom) 420 Wp PV pumping system for 40 families, Brazil

2.5.4. India

The largest number of solar pumps in one country is in India, where more than 1000 systems have been installed for village water supplies. Good responses have been reported along with wide user acceptance. The modules and systems have been indigenously designed and manufactured by Central Electronics Limited (CEL) in India. The potential for application of PV in India is so great that several other companies have started PV production including Bharat Heavy Electricals Limited (BHEL) and Rajasthan Electronics and Instruments Limited (REIL).

Most PV pumps installed have been for shallow wells. One recent study reported that in the north-east of India of 60 PV pumps in use in an irrigation project 58 had been working satisfactorily for more than 4 years.

3. SOLAR PUMPING TECHNOLOGY

3.1. Introduction to Photovoltaics

All PV systems can be divided up into three major components or subsystems. These are illustrated in figures 3.1. and 3.2. for village water supply and irrigation respectively, and are:

- The PV array which converts sunlight directly into electrical energy as d.c. electricity.

- The power conditioning (if any).

- The load (i.e., an electric motor coupled to a pump).

In the case of solar water pumping there is an additional component:

- The water storage and distribution system, which delivers the water to the point of use.

The cost of the PV modules is directly proportional to their power rating, and so it is important that all parts of the system are as efficient as possible. Losses in any component mean that more power and therefore a larger array is needed, which will push up the cost of the system. However, the borehole cost may actually exceed the pumping system cost in many remote locations, and for irrigation systems the field application system can cost anything from a few hundred to about $2500 per irrigated hectare.

With conventional gasoline or diesel fuelled pumps, storage is not so important because energy is stored in the fuel itself, and the pump can be started whenever water is needed. With solar pumping, water is not necessarily available on demand or in the required quantities. The day-to-day variations in solar irradiation mean that for some locations with successively cloudy days, it may be prudent to consider storage of a surplus of water pumped on sunny days for use on cloudy days.

Although PV systems may be perceived as 'high-tech' the only pieces of equipment from the above list which would be unfamiliar are the PV modules themselves. However, they also have the advantage that they are solid state (i.e., they have no moving parts), and so are extremely reliable, thus requiring no servicing at all except occasional cleaning to remove dust. All the other parts of the system, such as the motor/pump and distribution system could probably be repaired (if not manufactured) locally.

Figure 3.1: Components of a PV village water supply system

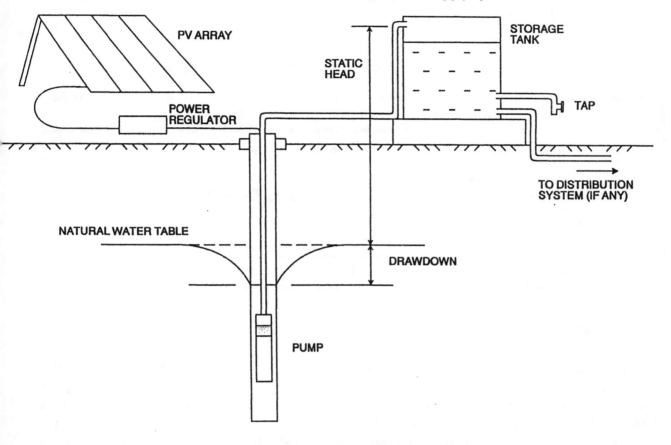

Figure 3.2: Components of a PV irrigation system

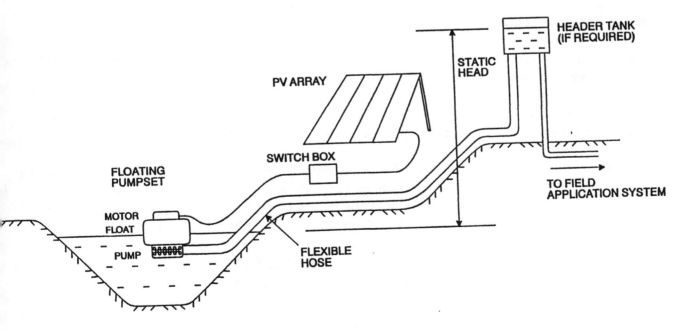

The advantages and disadvantages of PV in comparison to conventional pumping methods (i.e., diesel) are listed in table 3.1. below.

Table 3.1: Comparison of solar and conventional pumping systems

Advantages	Disadvantages
No fuel required	Relatively expensive for all but small applications
Little maintenance needed	Uses unfamiliar technology
Lifetime of modules very long (around 20 to 30 years)	Parts may be hard to obtain
Environmentally benign	

3.2. Photovoltaic arrays

The basic unit of PV equipment as far as the buyer is concerned is the PV module. These are the building blocks of the complete PV array and may be connected in series or parallel to give the required voltage and current characteristics for the load. Modules and arrays are rated in terms of the total power they can produce in Watts at a given radiation level (1000 W/m^2) and at a temperature of 25 C. This unit is the known as the peak-Watt (Wp). A typical module might measure 1 m X 0.4 m in area, and have a rating of 60 Wp. A typical array of modules for water pumping may be rated at anything from about a hundred Wp to a few kWp. A typical example is shown in figure 3.3., which shows a 600 Wp array in Brazil.

Because of the modular nature of PV arrays, their size is proportional to their cost, and so there are no economies of scale involved.

It is important to realise that the actual electrical power output of the array will be considerably less than the rated value. This is basically for two reasons:

(i)   The intensity of the sun will almost certainly be less than   1000 W/m^2, and thus the actual power output will be proportionately less. The solar intensity varie approximately as a cosine over the course of the day, and so the solar irradiance averaged over the hours of daylight will be about 500 W/m^2 in a good location.

(ii)   The impedance of the array must be perfectly matched to that of the load to perform at its full efficiency. This is called the maximum power point (MPP), and defines a certain voltage and current characteristic for a given solar intensity. Without specialised electronic equipment the array will never operate at exactly its MPP, although a well designed system should be fairly close.

Figure 3.3: PV array for a 600 Wp pumping system in Brazil

The array performance also varies with the module operating temperature, with efficiency decreasing at higher temperatures. Efficiency will drop by 0.5% fractionally per degree celsius increase in operating temperature. In the daytime the surface temperature of the modules can get as high as 60 C in the tropics, resulting in a 16% reduction in efficiency.

The result of the above effects is that the actual power output of a typical array will, over the course of the day, generally be less than half of the rated value.

Tracking arrays are available which operate either automatically or manually. The array support is designed such that the array follows the sun, thus boosting the daily energy capture efficiency, and allowing a smaller array rating to be used for a certain application. The automatic tracking array can either be operated electrically (active) or hydraulically (passive). The passive tracking arrays are the only ones usually considered reliable enough for use in developing countries. Manually movable arrays are of more use in PV irrigation situations, where the system is not likely to be left unattended for more than an hour or two. However, it is only in certain situations that the extra power gained from using a tracking array will offset its additional cost.

3.2.1. Solar cell technology

Most modern photovoltaic devices use silicon as the base material, mainly as mono-crystalline or multi-crystalline cells, but recently also in amorphous (non-crystalline) form.

A mono-crystalline cell is made from a thin wafer of a high purity silicon crystal, doped with a minute quantity of Boron. Phosphorous is diffused in the active surface of the wafer. When sunlight falls onto the cell a voltage is generated between the front and back surfaces. The front electrical contact is made by a metallic grid, and the back contact usually covers the whole surface. An anti-reflective coating is applied to the front surface.

Solar cells are interconnected in series and parallel to achieve the desired operating voltage and current. They are then protected by encapsulation between glass and a tough resin back. This is all held together by a steel or aluminium frame to form a module.

The efficiency of conversion from sunlight to electrical power is typically 12-15 % for mono-crystalline cells and around 10 % for the poly-crystalline. At the standard irradiance of 1000 W/m^2 a 100 mm diameter circular (or 100 mm square) cell will produce about 1 Watt of electrical power. The performance of PV cells does not appreciably decline with age, and trouble-free service can be expected for up to 20 or 30 years for a modern module.

Several manufacturers are pioneering the use of amorphous silicon PV modules. Although efficiencies are lower than for poly- or mono-crystalline cells, manufacturing costs per Wp can be lower as only a thin film of silicon is used. However, most amorphous cells tend to degrade somewhat with time, and are not yet widely accepted by the PV user community. There is a great deal of research on-going in this area, and they can be expected to play a more significant role in PV pumping during the coming decade. More promise may be shown by exotic thin film materials such as Copper Indium Diselenide and Cadmium Telluride, and increased capture efficiencies using multi-layer cells to absorb different wavebands.

22

3.3. Motors and pumps

3.3.1. General

In some cases it is feasible to utilise off-the-shelf, mass produced motors and pumps. However, some manufacturers have developed specialised pumps and motors with an above average efficiency to minimise overall system costs. The pump/motor subsystem operates in a different way to a conventional motor because the power supply varies as the incident solar energy changes. Most motors are designed for maximum efficiency at certain voltage/current characteristics, and performance can drop off quickly away from this operating point. In a solar powered system the motor/pump subsystem must be able to work fairly efficiently over a range of voltage and current levels. Although these specialised motors cost more than conventional motors, this is outweighed by the cost saving in terms of the number of PV modules required.

The usual measure of the motor performance is the power efficiency, which is the ratio of hydraulic output power to electrical input power. This is an instantaneous measurement, and is a maximum for the design conditions. The daily energy efficiency of the subsystem is a time average of the power efficiency. This depends on variations in the power efficiency, and thus on the solar irradiance profile for the day. The daily energy efficiency is more useful, as this determines the required array size for a given hydraulic duty. Consequently this determines, to a large extent, the cost of the pumping system.

The motor/pump requires a certain minimum power input to start working, and different pump types respond in different ways. A centrifugal pump will begin to rotate at very low light levels, but will not lift any water through the head until the pump reaches a certain speed. This power threshold increases with the required pumping head. A reciprocating positive displacement pump needs a very high torque to start it, as the pump is pushing against the whole head. Thus more current will be needed for starting than running. This can mean that without power conditioning the pump will not start until quite a time after sunrise because the maximum current from a PV cell is limited for a given irradiance level, and at high currents, the cell is far from its maximum power point (MPP). The starting and stopping threshold of a pump is an important parameter, because it determines how much of the total daily insolation is not used for useful pumping. A typical starting threshold might be 300 W/m^2. On overcast days the irradiance may not exceed this and the pump may not operate at all.

3.3.2. Motor technology

There are three types of motor commonly used for PV pumping applications, and each has its good and bad points. For this reason no one type has yet emerged as a standard, and the potential buyer should therefore be aware of the pros and cons of the motors in the various systems on offer.

- Brushed type permanent magnet d.c. motors
- Brushless permanent magnet d.c. motors
- A.c. motors

The immediate choice is between a.c. or d.c. motors. In terms of simplicity the d.c. motor is an attractive option because PV modules only produce direct current, and less specialised power conditioning equipment is needed. In many low power situations the array can be directly coupled to the motor with no electronics at all.

For higher power applications (>250 W) an a.c motor may be used in conjunction with an inverter (a device that turns d.c. power into a.c power). The range of a.c. motors available is much greater and the prices are generally lower. However the inverter is a relatively expensive piece of equipment and for small systems, the savings from using a cheaper a.c. motor may be offset by the additional cost of the inverter (around $1000). This configuration is therefore usually only the norm for larger systems or deep borehole applications. Commercially available inverters are designed to perform best at certain input and output characteristics, and the variable nature of the PV power supply may adversely affect their efficiency. It should also be noted that in a recent study of pump failures in Mali, 20 % were caused by inverter failure, the second largest cause after problems due to dirt.

The d.c. units used for PV applications are generally of the permanent magnet type. In a conventional d.c. motor the magnetic field is produced electromagnetically by the field windings. While more output power can be obtained in this way for a given motor size, valuable energy is consumed by the field windings. For PV driven motors a permanent magnet is used to produce the magnetic field and so no power is consumed in the field windings, leading to higher efficiencies. Smaller PV array sizes may therefore be used for the same hydraulic duty.

It is inherent in the design of d.c. motors that brushes are needed to transmit the power to the commutator. These are usually made of graphite, and so will wear down over a period of time and require replacement. A typical replacement interval for modern pumps is every two years (or 2000 to 4000 pumping hours). Although replacement is not difficult, it does entail removal of the pump/motor (for submerged units) and means that villagers must be trained in an extra aspect of maintenance. There will obviously also be increased maintenance costs associated with brushed motors.

In the brushless d.c. motor the permanent magnet is the central rotor, and the field windings are electronically switched by means of a rotor position sensor. The extra electronic circuitry adds cost and possible reliability problems to this choice of motor, but during the last few years designs have improved greatly. Many established manufacturers now use brushless motors as a matter of course, and this is likely to be a growing trend for small pumping applications in the future.

The pros and cons of different motor types are summarised in table 3.2. below:

Table 3.2: Features of various motor types

Motor Type	Advantages	Disadvantages
Brushed d.c.	Simple and efficient for small loads. No complex control circuitry needed.	Brushes need replacing periodically.
Brushless d.c.	Efficient. No maintenance required.	Electronic commutation adds extra expense and possible breakdown risk.
A.C. Motor	Larger range available for larger loads. Cheaper than d.c. motors.	Less efficient than d.c. units. Needs an inverter, adding extra cost and increased breakdown risk.

3.3.3. Pump technology

In the design of any pumping system, the pump itself is the most important component. Pumps can generally be divided into two categories, centrifugal and positive displacement. These two types of pump have inherently different characteristics and are suited to different operating conditions. When purchasing a pumping system, the supplier will offer a type of pump consistent will the pumping situation, but it is nevertheless important that the user should appreciate the various types available.

Centrifugal pumps are designed for fixed head applications and their water output increases in proportion to their speed of rotation. The principle of operation is that water enters at the centre of the pump and a rotating impeller throws water outwards due to centrifugal force. The water outlet is on the outside of the impeller cavity and thus a pressure difference is created between the inlet and the outlet of the pump.

A pump with just one impeller is called single stage, but most borehole pumps are multistage. This means that the outlet from one impeller feeds into the centre of another, each one adding a further pressure difference. When the speed of rotation gets high enough this pressure will be enough to lift water through the head and pumping will begin. Centrifugal pumps have an optimum efficiency at a certain design head and design rotation speed. At heads and flows away from the design point their efficiency decreases, and so it is important to accurately specify your requirements when ordering a system. However, because of their low starting torque, they offer the possibility of achieving a close natural match with a PV array over a broad range of operating conditions.

Centrifugal pumps are seldom used for suction lifts of more than 6 or 7 metres and are more reliably operated in submerged or floating motor/pump sets (in fact the theoretical absolute maximum for suction lift is about 9 m). This is because they are not inherently self-priming and can easily lose their prime at higher suction heads. However, submersible pumpsets may lift water from many tens of metres, depending on the number of stages and operating speed. An example of a typical multi-stage submersible centrifugal pump is shown in figure 3.4.

Positive displacement pumps have a water output which is almost independent of head, but directly proportional to speed. These pumps employ a piston/cylinder arrangement, or cavity of variable size, and so when the pump starts, water is forced against the entire head. Frictional forces are higher than in centrifugal pumps, because contact of moving surfaces is necessary to 'positively displace' the pumped fluid. At high heads and low speeds the frictional forces are small relative to the hydrostatic forces. Consequently for high heads displacement pumps may be the more efficient choice. At lower heads (less than about 15 m) the frictional forces are large compared to the hydrostatic forces and so efficiency is low and a displacement pump is less likely to be used.

A factor to consider when coupling a positive displacement pump to a PV array is the cyclical nature of the load on the motor. This causes variations in the electrical impedance of the load as seen by the PV array, and so the array will fluctuate around (and hence away from) its maximum power point. This is particularly a problem during the high torques experienced on starting (see section 3.3.1.). This means that electronic power conditioning is sometimes needed to smooth out these impedance changes by dynamically matching the array and motor impedances. Smoothing the motor torque can also be performed mechanically by the addition of a flywheel or counterweight.

These power matching problems are not experienced by centrifugal pumps, which exert a smooth, constant torque on their motor.

The most common type of positive displacement pump configuration is the Reciprocating Jack pump or 'nodding donkey', in which a submerged piston is driven by a vertical shaft. This is illustrated in figure 3.6c.

A less common positive displacement pump that has maintained a small market share since its introduction as a solar pump is the 'progressive cavity' or 'mono' pump. This is shown schematically in figure 3.5. and is composed of a helical rotor inside a specially shaped, flexible stator. A single helical cavity is formed between the two, and rotation of the rotor forces this to progress upwards carrying the trapped fluid with it. This design has the advantages of a positive displacement pump (i.e., efficiency at high heads) with the good electrical load characteristics of a centrifugal pump (smooth torque, constant impedance for a given speed and head). Manufacture of these pumps is not as straightforward as for piston pumps due to the precision needed in shaping the rotor, and

26

problems were initially experienced in achieving a good seal between rotor and stator in dirty conditions. However, as recent improvements in rotor and stator design have solved these problems and led to reduced starting torques, this type of pump could have a wide application in the future.

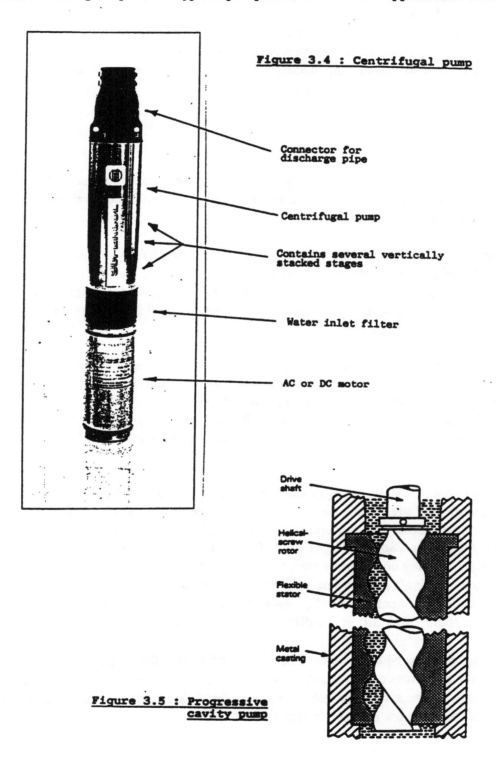

Figure 3.4 : Centrifugal pump

Connector for discharge pipe

Centrifugal pump

Contains several vertically stacked stages

Water inlet filter

AC or DC motor

Drive shaft

Helical-screw rotor

Flexible stator

Metal casting

Figure 3.5 : Progressive cavity pump

Readers may also come across a type of solar pump sometimes called an oscillation pump. This has been in use as a handpump for some time and is now marketed by Fluxinos of Italy as a solar pump, although usage is not yet wide enough to have obtained feedback from users. The device operates by storing energy in a series of compressible spheres within the pump, and releasing it in a pulsating fashion to carry water up the delivery pipe by its own inertia. Due to the non-constant forces on the motor, the pump is driven via a large flywheel. Potential advantages of this type of pump are that thee are no moving mechanical parts down-the-well, and in case of any array failure the device can be used as a handpump.

3.3.4. Subsystem configuration

There are several different system configurations that are suitable for use with solar power, and it is important to choose the right one for each application. The five main types are shown schematically in figure 3.6., and are:

(a) **Submerged motor/pump.** This is useful for medium depth (<50m) borehole applications using centrifugal pumps. Advantages are that it is easy to install with flexible pipework, and the motor-pumpset is submerged, away from potential damage. Motors may be d.c. (brushed or brushless) or a.c., although a.c. is most common, particularly for high power and deep borehole applications.

(b) **Surface motor/Submerged pump.** This design allows easy access to the motor for brush changes, but is now becoming increasingly unpopular for a number of reasons. Reliability tends to be poor when used with centrifugal pumps due to bearing wear, and installation is costly. There are also significant power losses from the shaft bearings due to vibration and friction. Data from a monitoring program in Mali has shown that most surface motor units are being replaced by submersible systems.

(c) **Jack pump (Reciprocating positive displacement pump).** These are also known as 'nodding donkey' pumps and are mainly used in very deep, low flow borehole applications. The motor is mounted above ground and features a balance weight to counter the cyclic force exerted on the motor by the pump. Some designs use different gear ratios for different parts of the cycle to improve the matching on the power stroke. Because the shaft does not rotate but moves vertically (hence 'reciprocating') it does not suffer the bearing-associated problems of design (b) above. Above ground components tend to be heavy and robust, and installation is expensive, both in terms of the motor/pump and also in the drilling of the deep borehole.

(d) **Floating motor-pump sets.** The portability of the floating pump set makes it ideal for irrigation pumping from canals and open wells. The pump moves with the water level, and so is not likely to run dry. The arrays are often mounted on wheels to allow easy movement. Obviously, this design is not suitable for borehole pumping.

Figure 3.6: Various motor/pumpset configuration

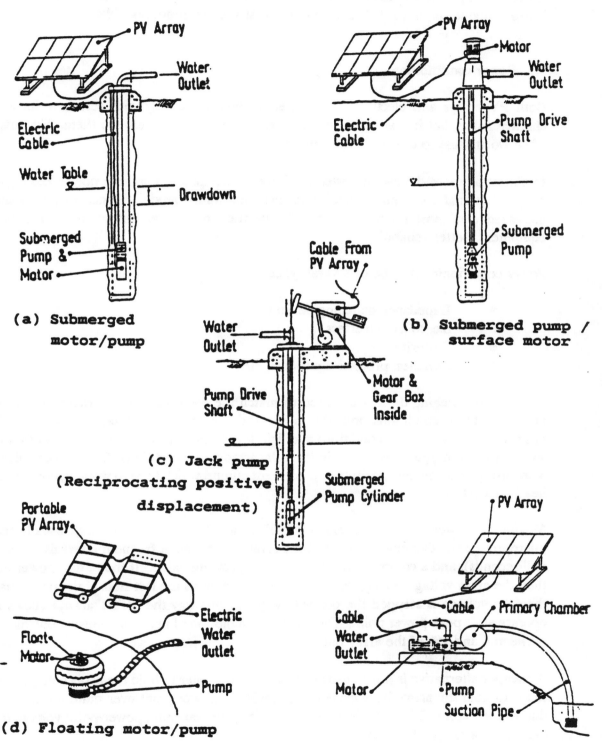

(a) Submerged motor/pump

(b) Submerged pump / surface motor

(c) Jack pump (Reciprocating positive displacement)

(d) Floating motor/pump

(e) Surface suction pumpset

(e)   **Surface suction pumpsets.** This configuration is not generally used unattended, due to self starting and priming problems, particularly at high suction heads. The physical limit on suction heads is about 8m, but it is better to operate at the minimum possible.

3.4. Power conditioning

Both PV arrays and electric motors operate most effectively with certain voltage and current characteristics, but it may be difficult to obtain a good match between them. For this reason it is often worthwhile to use some form of power conditioning.

In almost all cases the use of power conditioning equipment implies a power loss (typically 5% or less), additional cost and an additional possible failure mode. Hence to justify their use, the increased cost must be compensated for by the extra water output, or in the case of safety equipment, better reliability.

Power conditioning may be of several types:

- Impedance matching devices
- d.c. to a.c. inverters
- Batteries
- Switches, protective cut-outs, etc.

Impedance matching devices are used to match the electrical characteristics of the load and the array. In these conditions both the motor and the array can function close to their maximum efficiency over a range of conditions and light levels. For instance, high currents may be produced so that the motor/pump will start in low solar irradiance conditions. This is particularly important when using reciprocating positive displacement pumps, which have to work against the full pumping head on start-up.

Maximum power point trackers (MPPT's) are "intelligent" devices, usually employing a microprocessor. The power output of the array is sampled at frequent intervals (typically every 30 milliseconds), and a comparison made with the previous value. If the output power has increased then the array voltage is stepped in the same direction as the last step. If the power has decreased then the voltage is stepped the opposite way. In this way the MPPT always allows the array to operate at its peak power point. The power consumption of maximum power controllers is typically between 4 and 7% of the array output.

A cheaper alternative is just to use an electronic controller to hold the array voltage constant. This tends to hold the array fairly close to its maximum power point over quite a range of conditions, due to the nature of its electrical characteristics (the maximum power point voltage does not vary over quite a range of insolation levels).

Inverters convert direct current to alternating current, and are used to enable PV arrays to drive a.c. motors. Inverter efficiencies can be as high as 97%, but circuitry should involve some form of impedance matching for best results. Inverters tend to be rather expensive (>$1000), and so on small systems their additional cost may not compensate for the reduced cost of an a.c. motor. Reliability can also be a problem, with 20% of pumping system failures in a programme in Mali being due to inverter malfunction. Choosing an inverter is critical to its effectiveness in the system, as many units are designed for certain characteristics, and have a poor part-load efficiency. However, there are now several variable frequency inverters on the market that have been specially designed for PV applications. These have so far proved to be reliable and efficient (>95%) even on part load.

Batteries also provide a means of impedance matching. A battery can usefully store the energy from the array at irradiance levels too low to start a pump. Pumping can therefore start at low or even zero light levels as required. Because of their fixed voltage operation, designers can optimise the motor/pump subsystem for maximum efficiency. One drawback is that battery efficiency may be as low as 70% through self-discharge, and so this may offset the benefit gained at low irradiance. They also require regular maintenance and have a shorter operational life than the rest of the pumping system. Very few solar pumping systems include batteries at present, although research work continues in this field.

Switches and cut-outs: These protect components against power surges or damaging electrical conditions that may be caused by failure of other components, incorrect use or connection, or other possible malfunction. A system involving batteries may include a low-voltage cut-off to protect the batteries against deep discharge. If a pump runs dry, the motor may over-speed and burn out. Therefore if this is a possibility (i.e., with unattended borehole systems) a water level detector or over-speed cut-out device should be used.

3.5. Storage and distribution

3.5.1. General

The importance of an effective delivery system cannot be over-stressed, as its characteristics will have a significant effect of the size of pump required and the overall cost of the system. The main factors to consider are:

- The volumetric efficiency of storage and distribution. This is the fraction of the pumped water which actually reaches its point of use. This is particularly vital in irrigation applications, in which a great deal can be lost. This would mean a larger pump, a larger array and thus higher costs.

- The total pumped head. This is the sum of the static head (pressure needed to pump the water up to the level of the storage tank), the dynamic head (pressure loss due to friction and turbulence in the pipework) and any increase in head due to drawdown in the well or borehole.

- The actual cost of the storage and distribution system itself. For a large irrigation system this could be a significant part of the total system cost.

Often systems that are efficient in their use of water (such as pressurised-drip systems for irrigation) require high driving heads, and so to optimise the overall system performance, any increase in delivery efficiency should be weighed against the cost of increasing the driving head. However, there are exceptions to this relationship, and several high-efficiency ways do exist to distribute small quantities of irrigation water at low-heads, even using drip systems.

3.5.2. Water storage

With conventional pumping systems water is available on demand, and so short term storage is not such an important consideration. However, with solar pumping there may be significant variations in sunshine from day to day, and so water may not be available when it is most needed without some form of storage. The storage requirements for village supply and irrigation are rather different:

- For irrigation, artificial above ground storage is not feasible due to the large quantities of water involved, and would in fact, rarely be required. On a day to day basis, variations in supply are smoothed out by storage in the root zone of the crop. However, for larger systems some storage may be provided in the form of a shallow pond, which can be constructed very cheaply with local materials and labour. At certain times of year this could make a difference to the survival of a sensitive crop, and generally gives the farmer better control over his water management. Over-watering is not a problem with PV pumping, as the quantity of water delivered is unlikely ever to be sufficient to cause rotting. For systems with no storage, a small header tank may be appropriate to provide an even pressure to the distribution system (see figure 3.2.).

- For village water supplies some storage is essential, and should be adequate for several days water supply. Although 5 days may be desirable, in practice only about 2 or 3 days is usually affordable. A typical example would be a central raised storage tank close to the pump (to minimise dynamic head losses) and a piped distribution system to stand-pipes. Accepted water engineering practice, where cheap energy is available, is to position the tank high on a tower. However, PV pumps are typically used where either no distribution or only a few short branches to stand-pipes are required. The tank should be positioned high enough to provide only enough head necessary to carry water effectively through the distribution system, but it must be remembered that pump and array size is proportional to head,

32

and that a higher tank level will mean increased cost. For this reason tanks should be wide and squat rather than tall and thin, or the static head may significantly increase as the tank becomes full. Pipes, being a cheap part of the system should therefore be oversized to reduce dynamic losses to a minimum. Storage tanks for drinking water should always be covered to ensure a clean supply, and prevent entry of dirt, insects and animals. Tanks should preferably be lined to prevent losses through seepage and evaporation.

3.5.3. Water distribution systems

3.5.3.1. Irrigation

The distribution systems for a small scale irrigation scheme consists of (a) The conveyance network, to move water from the pump (or storage tank) to the field; (b) The field application method to apply water to the crops.

In many of the PV pumping systems installed to date there has been no 'new' application system as such, the PV system simply feeding into the existing network of channels. This is however, extremely wasteful, with losses up to and exceeding 50% of the water pumped. For a PV pumping system, where costs are directly proportional to output, losses of this magnitude are unacceptable, and it makes economic sense to install an appropriate distribution system to minimise the cost of the pump.

There are several common methods of application, and their suitability varies between different crops and different pumping systems. Their main characteristics are summarised below:

Channel Irrigation - Conveyance is in open channels, and although evaporation is low (typically 1%) there are very high losses due to seepage and weeds (30 to 50%). The static head depends on slope and length of channel, but for a small network for use with a PV pump, distances would be very short. Field application is in furrows, with most losses due to surface run-off and deep penetration below the root zone. Typical overall efficiency is only 40 to 60%, and this method is not advisable for use with PV systems.

Drip/Trickle Irrigation - Conveyance is by a main pipe and application is by many smaller perforated trickle pipes. Water losses can be very low (efficiencies around 85%) and with large pipes and low flow rates, dynamic head losses can be kept to a minimum. Traditional drip/trickle systems require a fairly high head to ensure even distribution. However, there are now a number of perforated or porous pipes available which can be operated for short (<100m) runs, under 3 to 4 metres of head. With flow to all parts of the field being continuous all through the daily hours, the rates of supply can be so low that quite small tubing (<125mm) can be used. A filter will usually be needed to prevent clogging, even when water is taken from wells. Special fabric filters with low head losses (50 cm) must be used (usually incorporated on the header box), as many filters made in the developed world incur dynamic head losses of up to 30m.

Hose and Basin Irrigation - Small basins (about 1m square) with the water led into them by flexible pipe or hose are potentially one of the most efficient of the traditional irrigation methods. Using the very simplest measuring techniques (i.e., a bucket) almost the exact quantity can be delivered to the basins by a person with a hose. When ample labour is available at almost zero cost this is frequently the most appropriate distribution system. A header tank with its supply level a metre or two above the ground, can supply a number of hoses, with diameters large enough that there is sufficient flow is by gravity. Volumetric efficiencies will naturally be extremely high, as all water is delivered exactly to the required place, and hand metered by the farm workers.

Sprinkle Irrigation - Although volumetric efficiency is high (typically 70%) this technique is not practical for solar pumps due to the high heads needed. For example, a coverage of diameter 6 to 35m, requires a head of 10 - 20 m.

Flood Irrigation - This method is hardly worth discussing in reference to PV pumps due to the enormous quantities of water required and the irregularity of demand. Fields are divided into basins filled within 10cm of bank tops. The basin size depends on available supply rate, but demand is uneven and size of pump determined by peak demand at flooding time. Used with paddy crops such as rice.

Table 3.3. below summarises the efficiencies, heads and suitability of each method:

Table 3.3: Efficiencies of irrigation application systems

Distribution Method	Application Efficiency	Typical Head	Suitability for Solar Pumping
Open Channels and furrows	40 - 60	0.5 - 1	Sometimes
Drip/Trickle	85	1 - 2	Yes
Hose/Basin	90	1 - 2	Yes
Sprinkler	70	10 - 20	No
Flood	40 - 50	0.5	No

From the above it is clear that there are really only two viable options for water distribution in PV irrigation systems: (i) Very simple hose and basin systems and (ii) Low-head drip or trickle systems. In both cases the emphasis is on low heads and high volumetric efficiencies. An important part of the site evaluation will be to collect the information required to provide estimates of the costs of the two alternative delivery systems. It may be found that the hose system will cost only about $250 per hectare, whereas the drip system will be closer to $2500 per hectare.

There may well be cropping circumstances which dictate the use of a drip system. Some high valued crops (e.g., strawberries) prefer to receive their moisture supplies from below their leaf-canopy, as serious burning of the leaves can occur in bright sunshine if they are watered from above.

Another important factor will be the labour supply. The hose system is clearly labour intensive, and as many as 5 people with hoses may be needed continuously to irrigate a hectare during the peak of the season. Nevertheless, if labour is cheap and plentiful and capital short then this may be the appropriate solution. On the other hand, the more even distribution from trickle irrigation will produce better yields, and may be justified by the resulting increase in revenue. Only full knowledge of local conditions can provide the answers to these questions.

3.5.3.2. Village water supply

In many cases there may be no distribution system, with villagers fetching their water from a tap at the storage tank. This reduces the head loss and reduces piping costs.

A prime factor to consider when deciding on a distribution system for livestock or village water supply, is the number of people or animals to be supplied by one pump. There are few economies of scale with solar pumping, and it can therefore be expected that several pumps are likely to have around the same cost as one pump with piped distribution. If individual well-yields are low this will be the limiting factor, but overall cost needs to be minimised for each application, and due allowance made for the cost of drilling extra boreholes or wells.

The number and positioning of stand-pipes should take into account the distance villagers must walk to fetch their water. The time used in collecting water must be weighed against the cost of additional piping.

3.6. Operation and performance

3.6.1. Subsystem efficiency

The bar graphs in figure 3.7. show typical 'average' and 'good' energy efficiency figures for different subsystem configurations with different pumping heads. These are defined for a daily solar irradiation of 5 kWh/m^2.

These efficiencies can only be taken as indicative, because they are dependent on irradiance and motor speed. The efficiency of conversion from the daily solar energy to hydraulic energy should be about 4 - 5 % for a good system.

3.6.2. Pumping performance

As it is usual to buy a complete system, it is more useful to have a feel for which systems perform best in which conditions, and over what range of conditions operation is possible. Figure 3.8. illustrates the typical ranges for different pump types as a function of output and head. The shading indicates the areas of the graph in which various configurations are most suitable, and can be used as a guide when performing a technical appraisal or life-cycle cost analysis as described in sections 6 and 7.

The axes of the graph are logarithmic. This is so that lines of constant volume-head product would be straight. If the performance range lines of the different pumps were drawn on a diagram of this kind, they would lie in roughly the same direction as the constant volume-head product lines, but be somewhat curved. This illustrates the fact that the actual volume-head product varies across the range of pumping conditions for a given pump, reaching a peak somewhere in the middle of the range.

Figure 3.7: Subsystem efficiencies: 'average' and 'good' cases

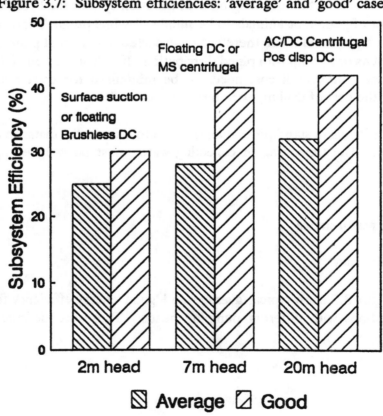

Figure 3.8: Configurations for various heads and volumes

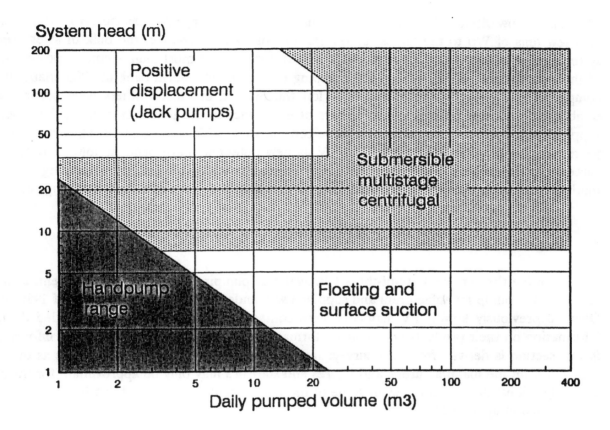

4. AVAILABLE EQUIPMENT

The market now offers PV pumping equipment to suit a wide range of scenarios, from the smallest of a few tens of Wp to the largest at several tens of kWp. However, the bulk of sales are for systems rated at less than 1500 Wp, and it is in this range that most manufacturers possess most of their experience. Since the last PV pumping update for the world bank in 1986, many of the companies and products then listed have left the market, but many others have become well established, and now possess valuable years of experience in the field. In general, the range of suppliers and equipment can be said to have consolidated: This is to be expected, as the past few years have been a formative period for PV as new potentials have been exploited, some more successfully than others. The coming decade promises an exciting future for PV pumping, as falling module costs bring the now established technology within the reach of a far wider range of users.

4.1. Range

To determine the current state-of-the-art in available pumping systems and equipment, a major survey of PV pump manufacturers and suppliers was undertaken during the winter of 1989/1990. Over 70 previously known PV suppliers were contacted, of whom about 20 provided detailed information on their products and their performance specifications. All the product information in this section is derived from this survey: Inclusions are therefore only as complete as the data received from the manufacturers, and figures and statistics may only be approximate for the sake of clarity. A list of the addresses of some of the major manufacturers and distributors contacted in contained in appendix C.

Most of the suppliers were able to offer surface suction and submersible pumps, with floating and jack pumps rather less common. Only Mono Pumps offer a borehole progressive cavity pump. Table 4.1 below summarises the capabilities of the suppliers that responded, detailing the configurations available together with their power ranges in Wp. In the table progressive cavity refers to submerged progressive cavity pumps with surface motor drives; this does not include the Chronar range of PC pumps which are of a 'surface suction' configuration.

The smallest units on offer are a 17 Wp domestic pump from Heliodinamica of Brazil and a 47 Wp submersible from Solarjack of the US. At the other end of the scale, the largest standard units are a 3.7 kWp submersible from Kyocera of Japan and a 5.1 kWp floating system from Total Energie of France. Very few maximum heads exceeded 120m with the highest being 300m for a Chronar jack-pump. The highest specified flow rate was 700 m³/day at 5m head for a Total Energie floating system.

Table 4.1: Configurations available and power range in Wp

SUPPLIER	SURFACE SUCTION	FLOATING	SUBMER-SIBLE	JACK	PROG. CAV.
BP Solar	yes 80-720	yes 320-480	yes 560-1400		
Chloride Solar		yes 320-640	yes 740-1480		
Chronar	yes 120-480			yes 160-960	
Dinh Co.	yes 100-800		yes 975-1625	yes 975-1625	
Duba	yes 120-200		yes 640-1280		
Fluxinos	Oscillation pump 360 Wp				
Grundfos			yes 250-1500		
Heliodinamica	yes 17-490		yes 735-1470		
Helios Technology	yes 270		yes 270-1800		
Hydrasol		yes 300-1400	yes 300-900		
KSB		yes 530	yes		
Kyocera			yes 50-3700	.	
McDonald	yes 130-1680		yes 320-1590		
Mono/Suntron	yes 120-1680				yes 120-1680
Siemens		yes 550-825	yes 770-1540		
Solarjack			yes 47-94	yes 106-424	
Total Energie		yes 240-5120	yes 640-3200		

4.2. Operational specifications

The manufacturers literature for all the pumping systems in the survey has been compiled into a common form in an attempt to provide a platform for comparison between systems. The full table of data is contained in appendix B. For each company there is an entry for each model of pumping system. If the system is available in many different configurations (i.e., different powers or heads) then some representative cases are included. For each entry the table details the pump type, array rating in Wp, the total system cost in US$ (FOB), the cost in $/Wp and the cost in $/m^4 at the design hydraulic duty. On the facing page the table continues with the specifications in terms of head (m), flow rate (m^3/day), and hydraulic duty (volume-head product, m^4). These are repeated for the minimum, mid-range and maximum specified heads. The column on the far right gives the reference solar insolation (in kWh/m^2/day) at which the pump specifications are defined.

The range of design performances for the dataset of commercially available systems is illustrated in figures 4.1 and 4.2. These show each system listed in appendix B as a point on a scatter diagram of pumped head (m) versus pumped volume (m^3/day). Each point represents the design operating conditions (mid-range head in the table in appendix B) and should therefore be near their maximum efficiency. Lines of constant hydraulic duty (volume-head product, m^4) have been added for reference. Figure 4.1 shows the whole dataset except for two very high flow systems (>300m^3/day) and one very high head system (130m). Figure 4.2 is a detail of the more commonly used region of this dataset. From these diagrams it can be seen that systems performing a duty of up to about 1000 m^4 are commonplace. By comparison with the scatter plot of field installations, figure 5.6 in section 5.1, it can be seen that the distribution of specifications of systems in use in the field matches that of the available products fairly well. The design head of the bulk of the available systems lies between 5 and 60 m, and the design daily pumped volume is generally less than 75 m^3/day.

4.3. System costs

4.3.1. History and nomenclature

During the last decade the real costs of PV pumping systems have fallen considerably, as has the range of costs. Figure 4.3 represents a summary of this decrease in prices and shows that although the fall is slowing down there is still a downward trend. As motor/pump technology is now well developed, the fall in prices reflects the continuously decreasing cost of PV modules, whose base level is currently around 5 $/Wp f.o.b. The PV array used to comprise the majority of the cost of a pumping system, but prices have now fallen to the extent that the array and pump are comparable components in terms of cost breakdown.

Figure 4.1: Pumped head vs volume scatter plot
(Commercial data: whole dataset)

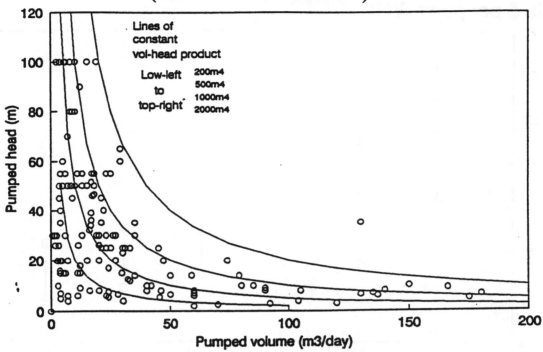

Figure 4.2: Pumped head vs volume scatter plot
(Commercial data: detail)

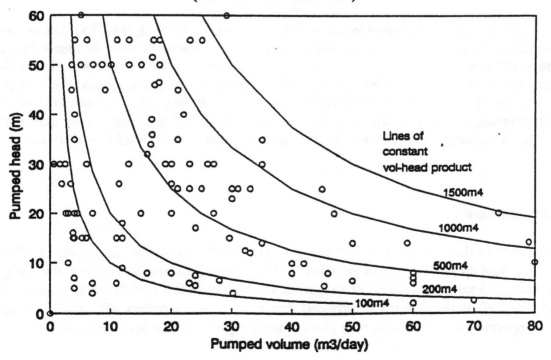

Figure 4.3: PV pump system price history

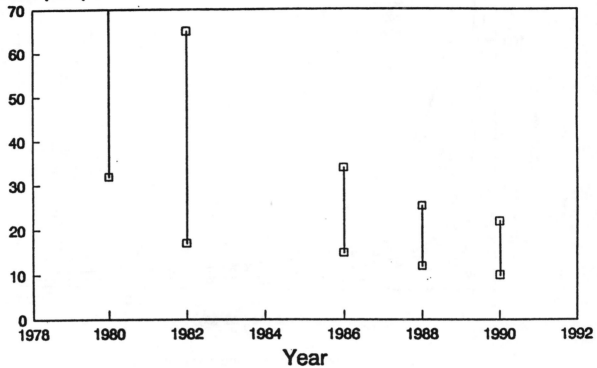

Just as module prices are described in terms of dollars per peak Watt ($/Wp), so it has become the norm to talk of PV pump system prices in the same units. However, this can be misleading, as the subsystem efficiency can vary greatly (see figure 3.7., section 3.6.1.) and will thus influence performance. A far more useful way of expressing prices is as cost per unit hydraulic duty ($/m⁴), as this is far more closely related to the actual performance of the pump. The two are numerically not that different, with the $/Wp value for a system usually being slightly larger than the $/m⁴ value. For the system costs reviewed in this section, the costs will usually be expressed in both forms.

4.3.2. Manufacturers data

The price data in this and the next sections were obtained in the winter and spring of 1989/1990 and are the most up to date figures available for PV pumping systems at the time of writing. The data represents a sample of over 150 systems of all types from the world's leading manufacturers and suppliers. However, it must be stressed that most of the suppliers will usually assess each proposal separately and have thus only given indicative prices to our survey request.

As it is normal to purchase a complete system rather than separate components, most of our attention in this section will be focused on system prices. We will examine these in several different ways, and in all cases prices are given as a range, within which the bulk of the systems lie. As there is a substantial difference between the costs of different system configurations, the price data have then been broken down into submersible, floating, surface suction and jack pumps.

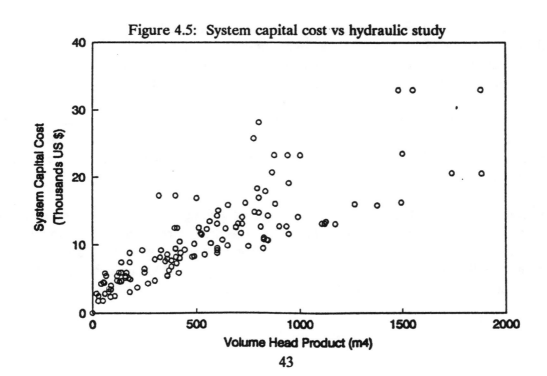

Figure 4.4: System capital cost vs array rating

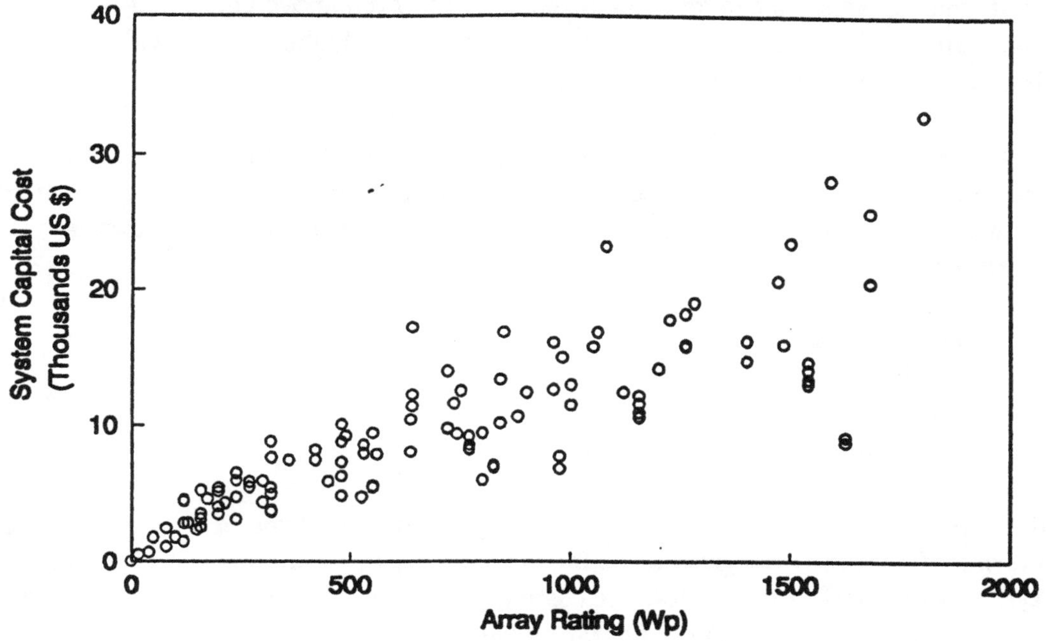

Figure 4.5: System capital cost vs hydraulic study

The first two graphs (figures 4.4 and 4.5) show the absolute system capital cost as a function of rated array power (Wp) and hydraulic duty (m^4) respectively for the whole dataset (i.e., all configurations). Both relationships are fairly well defined, but with greater scatter becoming apparent above 1000Wp (1000m^4).

Figures 4.6 and 4.7 show the same data but this time expressed in terms of $/Wp and $/$m^4$ respectively. Both graphs show a fair degree of scatter. This is to be expected, and is due to the varying prices and subsystem efficiencies of different configurations. In figure 4.5. prices range from 15-40 $/Wp at small array sizes and level out around 10-12 $/Wp above about 800 Wp. Figure 4.7 shows a large scatter at below 250 m^4, and a steady reduction at higher duties, with a well defined range minimum at about 12 $/$m^4$ above 700 m^4.

While the graphs described above are useful in giving an appreciation of the scope of the market, they are probably too general and scattered for system pricing estimates. In appendix E, are graphs of $/Wp against Wp and $/$m^4$ against m^4 in which the dataset has been broken down into the four main configurations:

- Surface suction
- Floating
- Submersible
- Jack pumps

In appendix E, figure E.1. (a to d) shows the 4 configurations in terms of $/Wp and figure E.2. (a to d) in terms of $/$m^4$. These graphs are far more useful in estimating the probable cost of a particular system when performing a life-cycle costing (as described in section 8.), as given a certain scenario it is likely that the required configuration will be known (see section 3.3.4.).

For quick reference the data contained in the figures in appendix E has been summarised in table 4.2(a and b). As smaller systems tend to be more expensive on a $/Wp or $/$m^4$/day basis, prices for two power ranges have been given for each pump configuration. In general a better correlation is found between $/$m^4$/day and hydraulic duty than between $/Wp and array Wp except for suction pumps. It is therefore preferable to use table 4.2b. where possible for submersibles, floating pumps and jack pumps, and table 4.2a. for surface suction pumps. However, remember that the system sizes are usually defined for an insolation of 6 kWh/m^2 in the manufacturers specifications, and so if the design month insolation is significantly below this it will be better to use the cost in terms of $/Wp for all the systems.

For each configuration and power range, a range of prices is given. This represents the approximate spread of the price data in the figures, and so the centre of this range should give you a realistic estimate of the capital cost. When calculating costs in this way, bear in mind that none of the systems retailed at lower than about $1500. Therefore use $1500 as a minimum figure and ignore any calculated system cost that works out below this.

Figure 4.6: System capital cost per Wp vs array rating

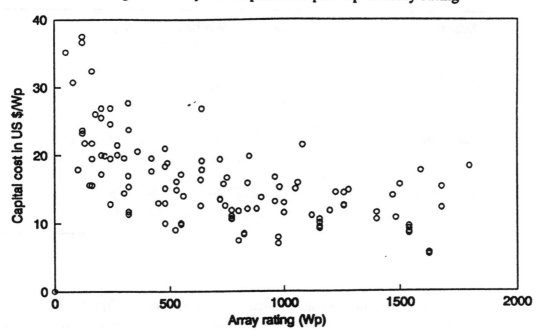

Figure 4.7: System capital cost per m⁴ vs hydraulic duty

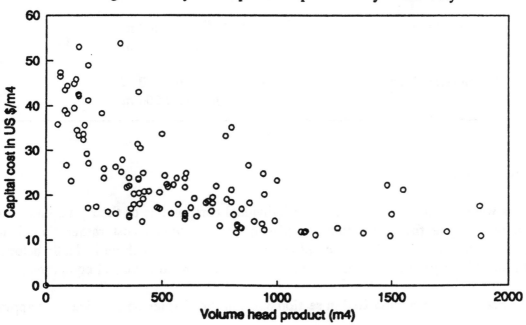

Table 4.2a: System price as a function of array rating in Wp

Submersibles	18 - 30 $/Wp below 400 Wp
	8 - 18 $/Wp above 400 Wp
Floating	11 - 17 $/Wp below 500 Wp
	8 - 13 $/Wp above 500 Wp
Jack Pumps	20 - 32 $/Wp below 500 Wp
	6 - 20 $/Wp above 500 Wp
Surface Suction	12 - 25 $/Wp below 300 Wp
	8 - 17 $/Wp above 300 Wp

Table 4.2b: System price as a function of hydraulic duty in m^4

Submersibles	20 - 50 $/m^4 below 400 m^4
	10 - 25 $/m^4 above 400 m^4
Floating	13 - 25 $/m^4 below 500 m^4
	8 - 12 $/m^4 above 500 m^4
Jack Pumps	18 - 30 $/m^4 below 400 m^4
	15 - 20 $/m^4 above 400 m^4
Surface Suction	25 - 55 $/m^4 below 200 m^4
	10 - 25 $/m^4 above 200 m^4

4.4. Component costs

Manufacturers were asked to provide prices of individual components of their pumping systems. Most of the replies give the prices that a replacement item would cost rather than being a breakdown of the system price. The different components are dealt with individually below, and are divided into PV arrays, motor/pumps and power conditioning and control equipment.

Component prices are also summarised in section 8.3.2. in the chapter on practical cost appraisal.

4.4.1. PV arrays

As this is typically a very long-lived component of the pumping system, not all of the systems suppliers give a price for replacement panels. However, many systems are expandable, and the addition of more modules is a simple procedure. As we have seen in previous sections, the base price to the consumer for PV modules is presently about 6 $/Wp, but those module prices listed tended to be between 10 and 13 $/Wp, with a mean of 11.5 $/Wp.

Array support structure, where separately quoted was between 1 and 2 $/Wp. Racks are available to hold 1, 4 or 8 modules. Tracking array supports were found to add about 4 $/Wp onto the system price.

4.4.2. Motor/pumpsets

Each configuration is dealt with separately below as there tends to be a considerable variation in prices.

Surface suction - The bulk of the prices ranged between $1000 and $1500 with a mean of $1330. This corresponds to between 2 and 4 $/Wp for the mid-range pumpsets, rising as high as 10 $/Wp for the smallest units (120 W). The cheapest unit was $675 for a low capacity pump from A Y MacDonald of the US.

Floating - There are very few floating pumps on the market (compared to surface suction or submersibles) and so the sample in this case is very small. However, those pumps that were available are priced fairly uniformly around $1000 for a small to medium sized unit, with one case of a high capacity unit at $2300.

Submersibles - Most of these motor/pumpsets lie in the range $1000 to $3000 with a mean of around $2300, which corresponds to about 1 to 3 $/Wp of rated array power. The smallest and cheapest unit is the Solarjack SDS at $425, designed for use with arrays as small as 47 Wp. Another small, low-cost unit is the Robby-24 from Helios at $702. In general, prices tend to vary more between different manufacturers than for different pumps within the size range of the same manufacturer.

Jack pumps - As with floating pumps, the sample is quite small, but shows that Jack pump prices vary in a well defined way with pump size, being between 10 and 15 $/Wp of the rated array power. As a lower boundary condition on this, no jack pumps were found costing below $3000. The mean price over the whole sample was $4870 with no units over $7500.

4.4.3. Power conditioning and control equipment

There is a large range of items that qualify as power control equipment, including inverters, maximum power point trackers and simple current controllers.

At the lower end of the price range, a simple electronic control module costs between $400 and $500, and will be sufficient for many small dc systems. Starter units (current boosters) are available for as little as $50 to $100.

At the higher power ranges ac motor/pumpsets with inverters become more common. Although there are many inverters on the market, there are relatively few that are suitable for use with photovoltaics (i.e., that can operate at high efficiency even at part load). The specialised inverters found in the survey tended to be priced at between $1000 and $2000 dollars for power rating up to about 1500W, although some higher power devices cost as much as $4000.

4.5. Specimen systems for typical loads

In an attempt to categorise the systems on the market for different applications, system suppliers were asked to specify a specimen system for 4 different pumping scenarios. The four scenarios defined were as follows, and were chosen to represent typical flow rates and heads likely to be required in various situations:

1) A capacity of 60 m^3/day at 2m head
2) A capacity of 60 m^3/day at 7m head
3) A capacity of 20 m^3/day at 20m head
4) A capacity of 40 m^3/day at 40m head

In all, 8 suppliers responded with details of appropriate systems, although only 7 of them could provide a system for the largest scenario, number 4. The details of the the responses from each manufacturer can be found in appendix D, but may be summarised as follows:

The first scenario of 60 m^3/day at 2m head corresponds to a micro-irrigation scheme or village water supply for about 1500 people. Pumping is likely to be from surface water (e.g., pond, canal or stream) or a shallow well, and the daily hydraulic duty is fairly small at 120 m^4. Three of the 8 replies used a floating motor/pumpset, 4 used surface suction pumps, and one did not specify a configuration. Array ratings ranged from 318 to 600 Wp with a mean of 480 Wp. System prices ranged from $5370 to $9340 with a mean of $7350.

The second scenario of 60 m^3/day at 7m head corresponds to a village water supply system for about 1500 people. Pumping is likely to be from a shallow well, and the daily hydraulic duty is around 420 m^4. Three suppliers specified floating centrifugal units, three specified surface suction pumps, one suggested a DC submersible and one gave no configuration data. Array ratings ranged from 600 to 745 Wp, with a mean size of 660 Wp. System prices ranged from $8030 to $11440 with a mean of $9290.

The third scenario of 20 m³/day at a head of 20 m is suitable for a small village of 500 people, corresponding to a hydraulic duty of 400 m⁴. At a depth of 20m, pumping would have to be from a borehole. Suppliers suggestions included 2 AC and 3 DC submersible motor/pumpsets, and three surface-motor driven mono pumps. The sizes of arrays specifies ranged between 420 and 840 with a mean of 580 Wp. There was a very great difference in system prices, ranging from $6841 to $16840 with a mean of $10340. The lower end of the range represents the surface-motor mono pups, while the most expensive are the dc submersibles.

The fourth and largest system defined was to pump 40 m³ at a head of 40m. The gives a fairly large hydraulic duty of 1600 m⁴, and corresponds to a borehole installation supplying a village of around 1000 people. Of the 7 systems suggested by suppliers, there were 3 ac and 1 dc submersible, and 3 mono type surface-motor shaft driven pumps. Two of the ac submersible systems used two pumps to achieve the requirement. Array ratings ranged between 1410 and 2968 Wp with a mean of 1841 Wp. System prices ranged from $16760 to $31930 with a mean of $24500.

One supplier (Suntron pty) offered a choice of tracking or static array for each scenario. The prices are identical for all the scenarios except number 4, in which the system with the tracking array is the cheaper by $3000.

4.6. Future trends in costs

Recent years have seen the cost (in real terms) of solar water pumping falling, largely due to reductions in the costs of PV modules. This trend is expected to continue in the next decade with the advancement of PV technology and more efficient manufacturing processes. PV module costs (direct from the manufacturer) are presently at about 5 $/Wp, and are predicted to have fallen to 3 $/Wp by about 1995. This 2 $/Wp decrease should be reflected in system prices, bringing the base level down to about 6 $/Wp. Although it should be feasible to produce modules at 1 $/Wp using thin film technologies, this is as yet some way off (i.e., post year 2000). It should also be borne in mind that previous PV price predictions have usually been over-optimistic, and that market forces as well as technical factors may influence future prices.

The costs of motor/pump units or inverters are not likely to decrease, although there should be some steady improvement in efficiencies and reliability.

5. GENERAL FIELD EXPERIENCE

Since the first solar pumps were installed in the field in the 1970s a great deal of experience has been gained by both manufacturers and operators. Solar pumping is now used in extremely diverse environments, from wealthy farms in the US to some of the worlds poorest communities in Africa and Asia. Although it is impossible to say exactly how many solar pumping systems have been installed around the world to date, the number was estimated at 6000 in 1988 and based on recent sales data has been estimated at close to 10,000 in 1990. The curve of cumulative pump sales in figure 5.1. rises almost exponentially, and is currently around the 2000 mark. Many of the earlier models have been replaced, and the number presently operating is thought be about 5,000, with 30 to 40% of these in the developing world. However, as pump sales accelerate and diversify, any attempt to objectively derive a total figure is certain to be an under-estimate.

The reasons for this increasing trend are essentially twofold:

(i)   The falling real prices of PV modules. This is due to both improving manufacturing technologies and price competition between the major module manufacturers.

(ii) Increasing awareness and acceptance by users. Solar pumping is now established as a reliable and cost effective technology by NGOs and governments in the developing world. PV is also becoming fashionable in the more affluent western countries, where its growth is stimulated by an ever increasing environmental awareness.

5.1. Characteristics of installations

The sizes of PV pumping systems and their uses vary enormously, The smallest consist of just 1 or 2 modules (about 80 Wp), sufficient to supply water for a just a few families. The smaller pumps are also widely used on ranches in the US, Canada and Australia for domestic supply and limited cattle watering. The largest systems in common use are rated at a few kWp, to supply drinking water for large villages or for irrigation.

By far the more prevalent use of PV pumping in the developing world is for village water supply and livestock watering. The large quantities of water and very seasonal demand make PV uneconomical for all but the smallest irrigation schemes (e.g., vegetable gardens or small area crops with a 'non-peaky' demand).

To gain some insight into the way PV pumping is used, we can look at several sets of statistics that divide up the installations according to different criteria:

Figure 5.1: World PV pump sales 1978 -1989

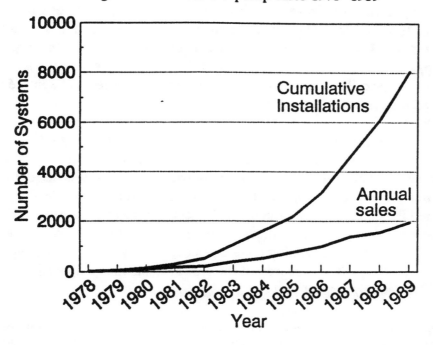

Figure 5.2: Breakdown of world market by application and remote market by region

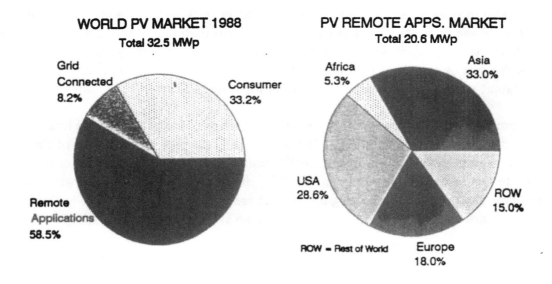

Beginning on a macro level, the pie-graphs in figure 5.2. reveal the distribution of the world PV remote market by application and by region. From figure 5.2.a. it can be seen that of total world PV module sales in 1988 of 32.5 MWp, remote power applications accounted for about 20 MWp (59%). It is believed that 8 to 10% of this remote market is used in water pumping. This amounts to 1.5 to 2.0 MWp/year. The regional breakdown of the remote power sector is shown in figure 5.2.b. The developing world PV pumping market is therefore expected to be in the region of 750 kWp per year (1988 figures).

The typical village water supply system in the developing world has an array rating of around 1000 Wp, and this is illustrated in the graphs in figures 5.3.a. and 5.3.b. This shows the distribution of sites as a percentage within several array rating ranges. The datasets used to compile this distribution contain all known PV pumping sites in Mali and Morocco respectively. Both countries have well established PV pumping programmes and are considered to be representative of pumping applications in the developing world. Although the distribution for Morocco peaks at a slightly lower value than for Mali, they both exhibit a main peak around systems of 1000 Wp, and a second smaller peak at about 5 kWp. These represent the typical small village application and the larger irrigation/large village supply respectively.

Village water supply is usually from boreholes (surface water should be avoided unless filtered due to probable contamination) with heads in the range 15 to 40m. Although solar pumps are theoretically capable of pumping over any head that a conventional pump can pump through, it is unusual to find them used with depths of greater than about 80m. The pumped volumes are typically between 20 and 40 m^3 per day. This corresponds to domestic water supply for villages of between 500 and 1000 people. There are only a handful in the sample providing greater than 120 m^3/day; these points are illustrated in figures 5.4. and 5.5., which show the distributions of pumped heads and daily water volumes from the Mali dataset. These are quite typical, and include pumping from diverse sources, including boreholes, wells and surface water.

It is more useful to work in terms of volume-head product (m^4) than hydraulic energy, as this bears more relation to the quantities used in water pumping calculations. Plotting a scatter diagram of pumped head vs daily pumped volume for a cross section of pumps (see figure 5.6.), it is found that the bulk of systems operate at a hydraulic duty of between 500 and 1000 m^4/day. The figure has lines of constant volume-head product marked on for reference.

Although firm figures do not exist, field experience has shown that practically all modern borehole pumps are of the electro-submersible multistage centrifugal type. The majority of these are AC, driven through an inverter. A few of the smaller units are brushless DC, driven either directly or through some power control device (e.g., MPPT). For shallow well or surface water, centrifugal floating pumps are the preferred configuration. These are exclusively DC and both brushed and brushless units are common. For deep-well work (>60m) the jackpump is still the only realistic option.

Figure 5.3: PV array size distribution (a) Mali (b) Morocco

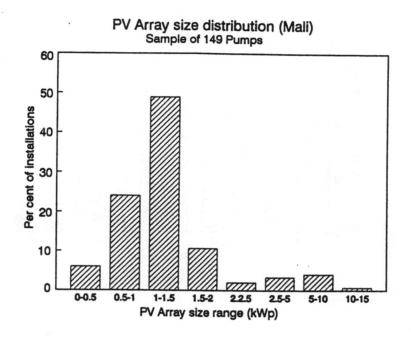

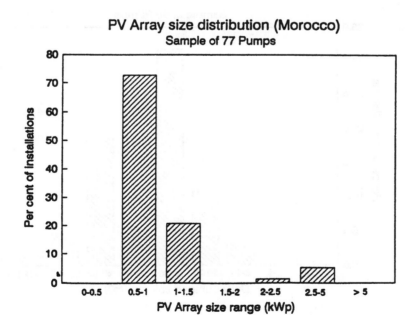

Figure 5.4: Pumped head distribution (Mali)

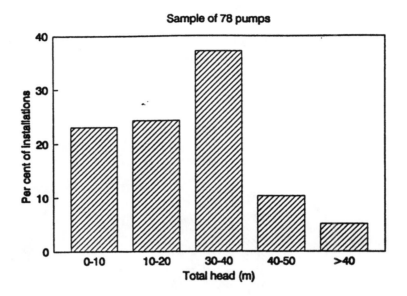

Figure 5.5: Daily pumped volume distribution (Mali)

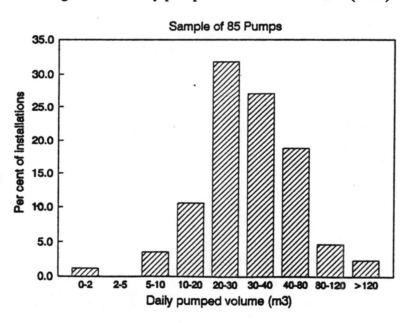

Figure 5.6: Pumped head vs daily volume for a sample of sites

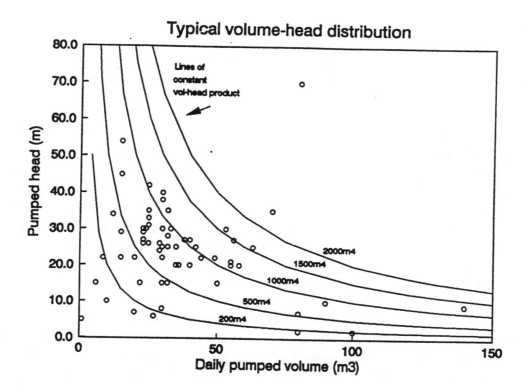

Due to the variable nature of renewable energy sources compared to conventional sources, at least some storage is a necessity for village water supply applications. However, the cost of water storage (i.e., concrete tank construction) is comparatively high, and as falling system prices make solar pumping increasingly favourable, the storage cost becomes a more significant fraction of the total cost. So although the desirable storage capacity for a village supply system may be five days, a tank of such a size is often not viable. Of a cross-section of 63 pumping systems examined, the mean storage capacity was only between a half and one days supply. Only one system had a capacity sufficient for more than 4 days storage. Tank volumes were found to be typically between 10 and 50 m³, with few over 150m³. Clearly, the storage volume is less critical if solar pumping is not the sole means of water supply (e.g., if a handpump exists as backup).

5.2. Performance

5.2.1. Definitions

The output of a solar pump varies over a day, and from day-to-day depending on the solar radiation received and the head the system has to pump through. It is therefore not useful to define the performance of the pump by its peak flow rate.

Similarly, the photovoltaic array power does not tell us what the performance will be, because the average daily efficiency of the subsystem (power conditioning, motor and pump) varies with system type and application. It is more convenient to rate solar pumps by their hydraulic energy equivalent, defined as the product of the volume pumped and the head it is pumped through. This is proportional to the useful hydraulic energy provided by the system and has units of $m.m^3 = m^4$. Clearly, the output of a given array will vary with the level of solar insolation, and so for a given reference insolation our system is defined by its volume-head product. For instance a system designed to pump 20 m^3/day through 15 m head in a location receiving an insolation of 5 kWh/m^2/day would be rated at 300 m^4/day at 5 kWh/m^2/day.

It is normal to characterise insolation figures by the mean number of kWh/m^2/day for the month, and therefore pumps are specified in terms of mean daily output. However, the relationship between irradiation and instantaneous hydraulic energy output may be slightly non-linear, and so the daily water output for a given system will differ somewhat from laboratory figures as variations occur in the shape of the solar radiation profile over the course of a day. For instance, continuous bright haze may produce a different output from brief spells of bright sunshine, even though the mean insolation in the two cases may be the same.

The nomogram in figure 6.2. can be used to determine the performance of a system in terms of daily volume-head product if the array rating, solar insolation and subsystem efficiency are known. This is essentially the reverse of the procedure for photovoltaic array sizing described in detail in the site
evaluation in section 6.5.

5.2.2. Performance in the field

In the previous section (5.2.1.) it was shown how the external factors (head, insolation, radiation profile) affect the performance of a solar pump of a given array rating. The other factor is internal to the system and is the subsystem daily energy efficiency. This is a very important parameter, as the lower the efficiency, the larger the array needed to do the same hydraulic duty, and thus the higher the system cost. The word 'subsystem' refers to the power conditioning, the motor and the pump, and the daily energy efficiency is the ratio of the total output hydraulic energy to the total electrical energy input from the array, taken over a whole day. This will be different to the instantaneous energy efficiency, which will vary as the day progresses, and will often be higher than the daily energy efficiency.

Both of the efficiencies defined above vary markedly with the configuration used. The most current figures for daily energy efficiencies are illustrated in the bar graph in figure 3.7. and discussed in sections 3.3.4. and 6.6.1. Naturally these are only approximate and will vary between different manufacturers and models. Figure 5.7. shows the way the total system efficiency (subsystem combined with array efficiency) has improved over the past decade. Modern systems should now fall within the 3.5 to 5.5 % range. Subsystem technology is now fairly well developed, and their efficiency improvements are only small compared to the advances being made in module efficiencies. This is reflected in the slight levelling out of the curve in figure 5.7. in recent years.

Figure 5.7: PV pumping system efficiency history
(mono-crystalline modules)

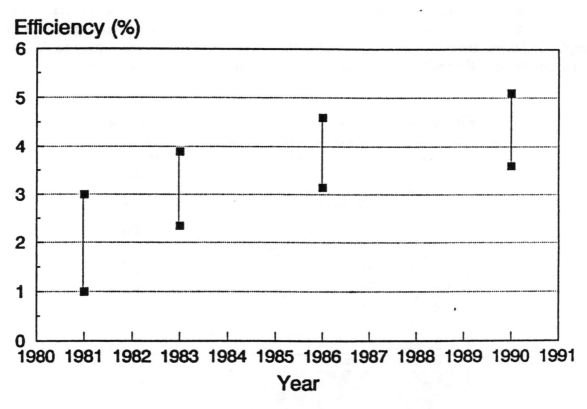

The subsystem efficiency will tend to decrease away from the design operating conditions. This is particularly true for centrifugal pumps, which should be carefully matched to the required head and flow. Figure 5.8. shows the way the operation of the subsystem for a centrifugal pump changes with changing conditions. The three boxes show a volume-head curve, a typical subsystem efficiency curve (sometimes called 'wire-to-water efficiency'),and the variation in electrical input power as a function of flow rate (source: Hydrasol Corporation). Reciprocating pumps tend to be most efficient at high heads, as the hydraulic forces become small compared to the frictional forces.

For a given solar insolation, the hydraulic duty (in terms of volume-head product) will be proportional to the array size and the daily energy efficiency. Thus by plotting the hydraulic duty against the array rating for different systems, the spread of the points gives us some idea of the variation in subsystem efficiencies. This has been done using the specifications for a selection of 55 pumps (various configurations) in figure 5.9. The pumps are all operating in one region of Africa (Mali) and should therefore be receiving a similar daily insolation (5.0 to 6.0 kWh/m²/day). As the design insolation is not exactly known for each site it would not be valid to calculate an efficiency value for each system, but the scatter of the points is indicative of their varying daily energy efficiencies. The best fit line corresponds to a daily subsystem efficiency of 40% if an insolation of 5 kWh/m² is assumed. As a rule of thumb, we can say for tropical regions that the hydraulic duty of a system (measured in m⁴) will be somewhat less than or equal to the array rating (measured in Wp). In figure 5.9. only 6 systems are above the one-to-one line that has been draw in. Detailed monitoring data has been accumulated for two currently operating pumps in this sample, and these are shown as squares rather than circles. Although they appear to the low end of the distribution, they still lie within the broad range of the specifications.

Figure 5.8: PV pump subsystem operational characteristics

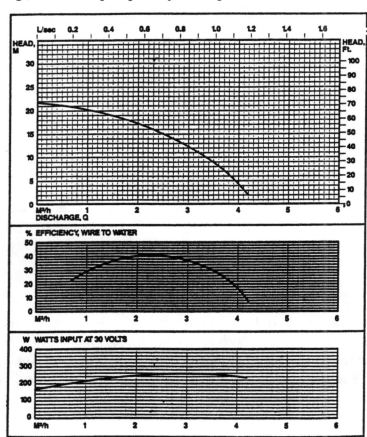

5.3. Reliability

One of the most significant areas of change in the solar pumping field has been the improvement in the reliability and robustness of systems. The past decade has enabled manufacturers to learn from previous failures, and there are now only a few problem areas remaining. The key to reliability in mechanical systems is simplicity. As new technological developments occur to improve pump performance or efficiency, they inevitably bring with them an additional possible failure mode, and the risk associated with new technology. As the new techniques become the norm, these problems are gradually ironed out. This has been seen very clearly in systems using a lot of electronics, such as inverters,
MPPTs or brushless DC motors. These technologies, although the cause of many early failures are now well established. Brushless DC motors are used routinely by many manufacturers in submersible and floating pumpsets, and there are several very robust inverters on the market that operate at above 95% efficiency even on part-load.

One significant factor given to the improved reliability has been the change from systems with surface mounted motors and submerged pumps to that of DC and AC submersible motor-pumpsets. Shaft and head-bearing maintenance was often a problem. Failures are almost always associated with motor, pump or power conditioning failures: Problems with the PV modules are very unusual.

One problem still not entirely overcome is that of pumps running dry. Some systems have been installed where the peak pump output at noon is greater than the borehole yield, such that the pump runs dry. This causes over-speed and burn-out of the motor unless a protection device is fitted. Falling water tables are also a possibility to be aware of.

Although statistics on reliability and failure modes are almost non-existent, some very useful data has been obtained from the pumping programme in Mali in West Africa. Of 126 pumps observed in 1988, only 6 were stopped. These were all surface-motor pumps awaiting replacement by submersibles due to design problems. Five pumps were abandoned due to wells running dry, and the remaining 112 were operational.

Sixty-six pumps were monitored from January 1983 to June 1989, during which time there were a total of 37 recorded failures. This corresponds to a mean of one failure in 139 pumping months, or a mean time between failures (MTBF) of over 30,000 hours. This compares very favourably with the 1500 hours MTBF for diesel systems and handpumps.

The breakdown of the failures was as follows:

- 7 inverters
- 4 motors
- 5 dc motor brushes
- 1 piping
- 6 wiring

- 1 due to dirt
- 3 miscellaneous

Figure 5.9: Hydraulic duty vs array rating for a sample of sites

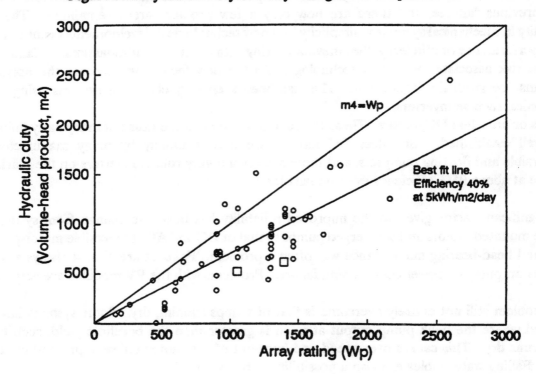

It must be remembered that the MTBF by itself is only a partial indicator of pump reliability; an equally important factor is the time needed to arrange and complete repair. This is the mean down time or MDT. From the above dataset, the MDT was 4 to 10 days, with a few cases over three months. Pumps were available for use for an average 99% of the time. This emphasises the need for a well organised support and maintenance service as well as careful choice and installation of the actual pumping systems. It should also be stressed that sufficient resources be allocated to obtaining accurate information on water resource characteristics.

5.4. Non-economic factors

The major areas of non-technical considerations relate to the financing , project design, implementation and infrastructure requirements of PV pumping programmes. It has been shown that involvement of the users in all stages of the project (including contributions to the cost) are essential for successful programmes.

In the past there has been some concern over introducing unfamiliar technology such as photovoltaics into rural areas, but experience has shown that acceptance by the local population is very high. The systems are accepted with great enthusiasm and are often a source of pride as they are demonstrated to visitors.

It has been found that the financial involvement of the community is a very important factor in project planning. The philosophy of donating the entire system and maintaining it free of charge has resulted in a lack of interest from the community, who often abandon the pump and return to their old water lifting methods when the pump fails. Most organisations now charge the villagers some nominal percentage of the capital cost, and the community pay all or most of the annual running costs. This approach has met with the approval of the beneficiary communities who collect payments at a village typically through a 'water committee' of two or three people. Charges are either made by volume used or by a set amount per family per year.

Infrastructure requirements such as community training, spare parts availability, preventative maintenance, fault-finding and repair services need to be carefully considered.

6. SITE AND SYSTEM EVALUATION FOR VILLAGE WATER SUPPLY

The end result of a site evaluation is to answer the question "What type and size of solar pump is best suited to the site, and what will its capabilities be?" for a given situation. This must include all aspects from the assessment of water demand and resource availability, to sizing of the various components of the pumping system. This chapter aims to give a practical means of estimating the various system parameters for a given scenario. All manufacturers will use their own sizing methods that are suited to their particular products. Therefore this is simply a methodology for an approximate sizing, to give the user some idea of the feasibility of solar pumping for his situation and of what to expect in terms of hardware requirements.

The requirements of village water supply and irrigation pumping are very different, as are the limiting factors in each case. It is therefore necessary to deal with site evaluation and sizing in very different ways, and so this chapter covers village water supply while the following one (section 7) covers irrigation.

There are three main technical factors that act as boundary conditions to the problem of site evaluation, and of which we must obtain realistic estimates. These are:

(i)   The demand for water
(ii)   The availability of water
(iii) The solar resource

Although the solar resource can be quite reliably defined, the assessment of thedemand for water and the availability of that water can present serious difficulties. Factors such as seasonally changing village populations and water tables and the relationship between well drawdown and pumping rate introduce unknown quantities that will complicate what at first glance may seem a simple problem. In addition, experience in West Africa has shown that following installation of a solar pump populations can double almost overnight.

The concept of demand is not always valid at all, as in many situations the limiting factor will be the production capacity of the well. In this case we must simply try to estimate what supply can be provided from such a well and to decide if this still justifies installation of a solar pump.

In this section a methodology will be presented that attempts to simplify this problem into a series of straight-forward steps, while still maintaining some degree of realism. The steps are shown in order in table 6.1. and will be dealt with in detail in the following sub-sections.

Clearly, for many communities or donors the question of cost will also be critical, and will itself be a factor affecting demand. For example, a lower per capita water supply may be acceptable if costs of a larger pumping system are too great. Therefore a certain amount of iteration may be necessary between this section and section 8 dealing with costs.

Table 6.1: Steps in village supply site evaluation methodology

I. From a cursory look at the site, determine the rough physical layout of the system and the likely water source.

II. Obtain information on wells and water resources. I.e., well diameters, yields, drawdowns and monthly groundwater information.

III. Using the available hydrological information, estimate the PV system size required to pump the maximum sustainable discharge from the well.

IV. Gather monthly data on the availability of solar energy in terms of average $kWh/m^2/day$.

V. Estimate the likely daily 'demand' for water by the village for each month.

VI. Calculate the array size that, for each month, would be necessary to provide the estimated demand found in V, assuming maximum well drawdown.

III. Choose the largest array size calculated in VI above, that is consistent with the maximum rating calculated in III.

VIII. Use the chosen array size to calculate the minimum capacity of water supplied per month (i.e., at maximum drawdown). Make a qualitative judgement on the likely mean monthly supply and the value of this service.

6.1. Physical system layout

At this point consideration must be given to the physical layout of the system. This will be dictated by the water source and the layout of the village. For village water supply, the water source will almost certainly be a dug well or borehole (the methodology described in this chapter is equally valid for either). Surface water is not usually considered fit to drink and would in any case not warrant a solar pump for the small quantities involved in village supply. To this end, determine the most logical positions of the borehole or well and the storage tank (if any) in relation to the point of use, and find a rough estimate for the lengths of piping that will be necessary. For a medium depth borehole the appropriate pump type is a submersible centrifugal pump.

The pipework will be probably be procured locally, and so the pipe diameter and tank size will be dependent on local availability. Typical pipe diameters may be 50 to 150 mm. If the pipes are too narrow the increased dynamic head will mean that a larger array is needed (this is the additional head requirement due to friction and turbulence within the pipework). As plastic pipe is cheaper than PV modules, pipes should be oversized such that head loss is minimal. This must be checked using the graph in figure 6.3. at a later stage in the site evaluation when flow rates are known.

The usual configuration for a village water supply network would be to place a storage tank close to the pump. Thus when calculating the total pumped head we only consider the pipework as far as the tank. This is because it is the tank and not the pump that provides the head for the distribution system (if included). The height of the tank does not need to be very large (1 or 2 m), although if a large distribution/stand-pipe system is to be used it may be worth calculating the expected head loss. The size of the tank is determined by how many days storage are wanted. Around five days is desirable if there is no other source of water, but one to two days is likely to be more common. Hence at a later stage in the calculation, when the average daily supply has been found, the tank volume is found by just multiplying the daily demand by the number of days storage capacity.

For a small village a distribution system is a needless expense and a simple tap (or taps) on the tank may be sufficient. The questions of storage and distribution are discussed in more detail in section 3.5.

6.2. Groundwater resources

The most likely limiting factor on the amount of water that can be pumped will be the availability and depth of groundwater. Most boreholes or wells drilled for village water supply have only enough capacity to use a handpump (<2 m^3/per hour) because they cannot refill fast enough. As water is pumped from a borehole, the water level will drop below that of the surrounding water table. It is the resulting head difference, known as the drawdown, that causes water to flow into the well though its walls. As the pumping rate is increased the drawdown also increases, until a point is reached where the well is emptied out. A lesser pumping rate, which can be sustained, must therefore be selected, such that the in-flow through the walls equals the outflow, with several metres of water still in the bottom of the well. This is the maximum sustainable pumping rate and will be used as an upper limit for design purposes. The exact relationship between the drawdown and the pumping rate will depend on the diameter and total depth of the borehole and the permeability of the rock or soil. The way the water table is 'drawn-down' around a borehole is illustrated in the schematic village water supply diagram in figure 3.1.

In situations where the water table is found in shallow sandy soil, say near a river or lake, the water is essentially submerged surface water. The abundant supply and highly permeable soil will mean that drawdown will be relatively small in these cases.

In areas where boreholes must be sunk into deep aquifers below caps of impermeable rock, the drawdown will be very much larger for the same pumping rate, and may form a significant part of the total head. Thus without including drawdown in the analysis there is a danger of under-sizing the pump, as the real water level in the well, once pumping begins, may be well below the static water table.

Even more serious is the danger of over-sizing the pump. If during the sunniest part of the day the pumping rate is too high the water level may drop below the level of the pump intake and the pump will free-run. This may cause the motor of the pump to overheat and burn out in quite a short time, although some manufacturers provide a cut-out to prevent dry-running.

Although it is never possible to exactly predict the drawdown relationship before drilling, some idea of its extent can be gained from data from neighbouring wells, or, if an existing well is being used, from the driller's records. When a borehole is drilled, a test should be performed by using a diesel pump to pump out water at certain rate and measure the drawdown when a steady state has been reached. The maximum safe rate and the corresponding drawdown should be recorded.

Seasonal changes in the depth of the static water table must also be taken into account. They will be more pronounced in submerged surface water situations, and depths can vary by up to 10m in some places. The level in deep aquifers tends to remain more constant. For practical purposes monthly data are desirable.

The minimum information needed to proceed with the site evaluation is shown in the example below:

Depth:	55 m
Inside diameter	6 inches
Test pumping rate	6.5 m³/hr
Static water level	14.0 m
Level at end of 8 hrs pumping	37.0 m

Depth to static water level:

Jan 19.0	Feb 19.5	Mar 20.0	Apr 21.0	May 18.5	Jun 16.0
Jul 14.0	Aug 13.5	Sep 14.0	Oct 15.5	Nov 17.0	Dec 18.0

In this example the drawdown at the test rate is therefore 37.0 - 14.0 = 16.0 m.

As records or test results for the site or well in question are rarely available in reality, it is usually necessary to examine data from nearby typical wells that have been measured regularly over a number of years. UNDP and other groundwater studies have now been conducted in most parts of the world and it is very rare to find a country with no groundwater survey reports at all.

For safety, the pump should be hung below the maximum drawdown level in the month with the lowest static water level, with a few metres leeway for annual anomalies in the level. (Note that the pumped head is measured from the water level in the well and does not depend on the depth of the pump below that level). The pump should not be too close to the bottom of the borehole as silt may build up there and clog it.

Knowing the maximum safe sustainable pumping rate, the drawdown below the static level at this rate, and the monthly static water depths we can calculate a rough maximum for the array size in the following way: The maximum depth to which the level should drop should be lower than the drawdown plus the deepest static level at the maximum safe pumping rate.

In the example:

Max drawdown: 23 m at a rate of 6.5 m³/hour
Lowest static: 20 m
So deepest water level 43 m
And the maximum total pumped head will be this plus the height of the tank (say 2m): total 45m.

Hence we can define a maximum safe hydraulic energy, and by estimating an efficiency for the motor/pump subsystem, a maximum safe electrical power. Assuming that at maximum sunshine intensity of 1000 W/m² the array produces its rated output then we can define a maximum safe array rating. We then know that an array of this size or less can never pump dry the well. This calculation is easily performed graphically with the aid of the nomogram in figure 6.1.: Convert the maximum pumping rate in m³/hr to litres per second by dividing it by 3.6 (i.e., 1 l/s equals 3.6 m³/hr). In the example this is 1.8 litres/sec. Finding this figure on the upper vertical axis trace across to the curve representing the appropriate total pumped head (interpolating between the 'head' lines if necessary). Then trace downwards to the appropriate subsystem efficiency curve. Use 30% efficiency for a typical multistage centrifugal pump. Finally trace across to the lower vertical axis to read off the corresponding array rating. In subsequent steps this will be taken as an upper limit on the array size. Using the example peak flow of 1.8 l/s with the example maximum pumped head of 45m gives a maximum array size of greater than 2500 Wp. This is therefore unlikely to be a limiting factor in our example.

6.3. The solar resource

The daily energy that a PV array is capable of producing is dependent on the intensity of the sun throughout the day and on the size of the array capturing that energy. Hence before we can calculate appropriate array sizes we need to know about the availability of solar energy at the site in question. The quantity in which we are interested is a measure of the total solar energy over a whole day. This is called the insolation and is usually expressed as a daily energy per unit area. For convenience we express the daily solar insolation in terms of kiloWatt hours per square metre (kWh/m^2). The insolation is a quite well defined quantity, and will not be greatly subject to local variations.

Ideally, month by month solar radiation data are required in order to properly assess the suitability of a site for a solar pump. It is not sufficient to size a pump on the basis of annual solar data, as sufficient water may not be provided in months of low solar insolation. If possible, data should be obtained from the nearest meteorological station, and allowance made for any known local variations in sunshine. A local University where solar research is in progress would be the next place to look.

It is probable that no such data will be available, and in this case a reasonable estimate can be obtained from the global radiation maps in appendix A. These show the total solar energy falling on a surface tilted at latitude angle plus 15° for each season.

67

Figure 6.1: Nomogram for peak flow rate calculation

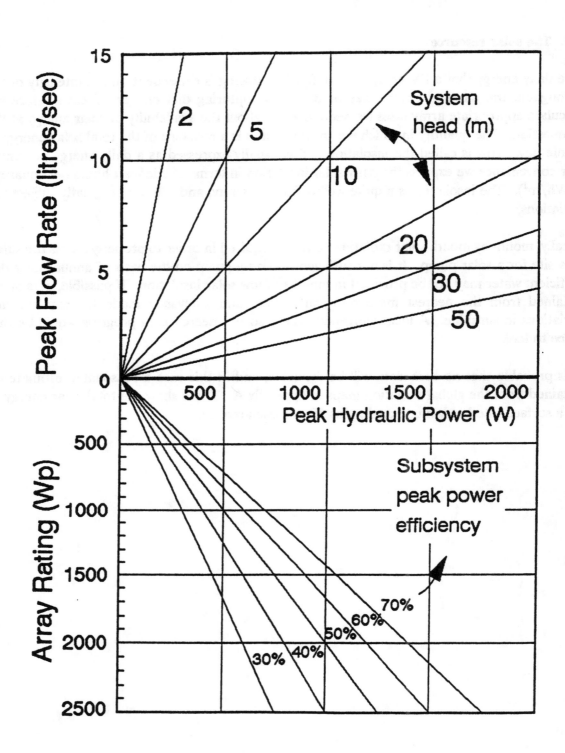

In theory an array tilted at the latitude angle gives the best mean annual solar energy collection. However, in practice it is better to tilt the array below this by several degrees to improve collection at lower sun angles. In addition, arrays should always be slightly tilted to allow rainwater to run off and keep the modules clean.

The seasons mentioned in the titles for each insolation chart in appendix A are those for the northern hemisphere. In the southern hemisphere the seasons will be reversed. The tilt angle is defined as the angle at which the array is raised from the horizontal, and is measured with the array pointing south in the northern hemisphere and north in the southern hemisphere.

6.4. Water requirements

Domestic water requirements vary markedly in response to the actual quantity of water available. For example, the average domestic consumption in western europe and the USA is between 100 and 150 litres per day. At the other end of the scale the level in rural areas of the developing world varies between about 5 and 35 litres per day. In severe conditions many people survive near the biological minimum of 2 litres per day.

Although precision in this matter is not possible, some figures are needed for sizing purposes, and the methods for obtaining them are discussed in this section.

Community population statistics are seldom recent or accurate, and populations change over time and with the seasons. This is particularly a problem in rural Africa: In Latin America and Asia populations and their demands tend to be more stable. In addition, per capita requirements change from season to season, and during the dry season it may be necessary to make provision for livestock watering. Some local knowledge of the population and their way of life is therefore an essential requirement in estimating the water demand for a village.

When sizing a solar pump it is important to remain flexible concerning the water demand. If a pump is too expensive, or if in just one month the pump cannot match the criteria set, then people will usually adapt to the conditions for a short period; this will be more preferable to them than having no pump at all. Similarly, if more water is being pumped than required for a certain month, they will be happy to use it. The most likely constraint on the available supply will be maximum extraction rate of the well, and at some times of year it may be necessary to restrict either the per capita use or the number of persons using the well. However, at this point we will calculate the ideal demand, as we will not know until we have sized the array if it exceeds the maximum safe size we defined in section 6.2.

A WHO survey in 1970 showed that the average per capita water consumption in developing countries ranges from 35 to 90 litres/day, with houses with a piped supply using up to five times more than if water must be carried from a public water point. The WHO subsequently defined 40 litres/day as a short term goal for the developing world. However, in many parts of Africa 10 l/day is regarded as an acceptable and realistic quantity for rural areas. This covers drinking water and cooking needs.

Low tech mechanical systems may come in the range of 20 to 25 litres per capita per day, and it is recommended that this figure be roughly used in calculating solar pumping requirements if local requirements are not directly known.

Seasonal changes in per capita consumption may be about 15 % either side of the mean, with the maximum being in the dry season. If no local information on water use is available this can be used in the monthly demand calculation and should be varied smoothly between the seasonal extremes.

Seasonal changes in village population may be more marked. As other water sources dry up, the people that were using them may come from further afield to use the solar pump. A survey of handpumps in a region of Ghana found that the average number of people using the pumps increased from 750 to 1250 in the dry season. Again, local knowledge is a great help in making assumptions of this kind.

Table 6.2. shows typical per capita daily water requirements for a range of livestock:

Table 6.2: Typical per capita daily water requirements for a range of livestock

ANIMAL	WATER REQUIREMENT (Litres per day)
Horses	50
Dairy Cattle	40
Steers	20
Pigs	20
Sheep	5
Goat	5
Poultry	0.1

The information detailed above can be collected into a table giving the minimum daily village requirements for each month. An example is shown in table 6.3. The first column shows the number of people needing water each month, and the second column shows their per capita requirement in litres. The third column is the ideal total daily demand in m³/day and is simply the first two columns multiplied together and divided by 1000 (to convert from litres to m³). In the example, April is the driest month and August the wettest. We should also account for any expected losses of water in the distribution and storage system. However, volumetric efficiencies for village water supply systems should be very high (above 95%) and so water losses can probably be ignored. This total demand is used in the following section in the array sizing calculation.

Table 6.3: Calculation of ideal daily
requirements for village water supply

Month	Number of people	Per capita demand (1)	Total daily demand m³
Jan	550	26	14
Feb	575	27	15
Mar	600	28	17
Apr	625	29	18
May	550	27	15
Jun	500	25	13
Jul	425	23	10
Aug	375	21	8
Sep	400	22	9
Oct	425	23	10
Nov	450	24	11
Dec	500	25	13

6.5. Array sizing

The size of the array and pump depends on the daily solar insolation and hydraulic energy requirement, which we shall express as the volume-head product. This is simply the daily pumped volume multiplied by the total pumped head. This must be calculated for each month, and the month that requires the largest array size is called the design month. This is so called because it is the 'worst-case' and represents the extreme conditions that we must design for. If the pump can meet the requirements in this month, then it can, by definition, meet them in every other month.

This maximum array size will, of course, be subject to the maximum safe size calculated in section 6.2.

The parameters needed to calculate the array size for each month have mainly been found in the previous sub-sections, and are as follows:

- Total pumped head
- Daily water requirement
- Daily solar insolation

A logical way to proceed is to create a table such as the example in table 6.4., with the months listed in the far left column. The following paragraphs will detail the filling in of each successive column.

Total pumped head (in metres) should be placed in the first column of the table. This consists of four components:

- The height of the tank (if any) above the surface
- The Depth to the static water table for each month
- Increase in water depth due to drawdown
- Additional dynamic head due to friction losses

The tank height should have been decided in the discussion on basic system layout in section 6.1. For our example we will use 2.0 m.

The depth to the static water table was found as part of the well statistics in section 6.2. Use monthly data if available.

The increase in water depth due to drawdown will not be known for a specific pumping rate, but we know the maximum allowable drawdown from the well statistics in section 6.2. For the sake of safety, this is the drawdown value we will use for all months. In the example this is 23m.

The dynamic head loss in the pipework from the pump to the tank is found from the graph in figure 6.3. (method described below under 'pipework sizing'). This should be minimal, not exceeding a few percent of the total pump head. For this stage of the calculation we can assume it to be zero.

The total head as experienced by the pump is thus the sum of the four components discussed above, and this sum should be added into the first column of the table for each month as in table 6.4.

Into the next column in table 6.4. should be entered the ideal daily water requirement, in m³/day, as was calculated in section 6.4. (in example table 6.3.).

72

In the third column we must calculate the daily volume-head product for each month, as this is proportional to the hydraulic energy required. For each month simply multiply together the total pumped head from column 1, and the daily water requirement from column two, and enter this in column 3. The units of the volume-head product are m^4.

Into the fourth column copy the values of daily solar insolation for each month (in $kWh/m^2/day$) that were found in section 6.3.

One final item that we need is the daily energy efficiency of the motor-pump subsystem. In the general layout discussion in section 6.1., the type of pump should have been decided upon, and the typical efficiencies for different configurations can be seen in figure 3.7. For most village water supply applications a borehole submersible will be most appropriate, and so a daily energy efficiency of around 30% will be used for the example.

We now have all the information necessary to calculate the array size that would be needed for each month's requirement. For the sake of simplicity, this calculation has been represented in graphical form as a nomogram (see figure 6.2.) as well as an equation.

Figure 6.2: Nonogram for PV array sizing

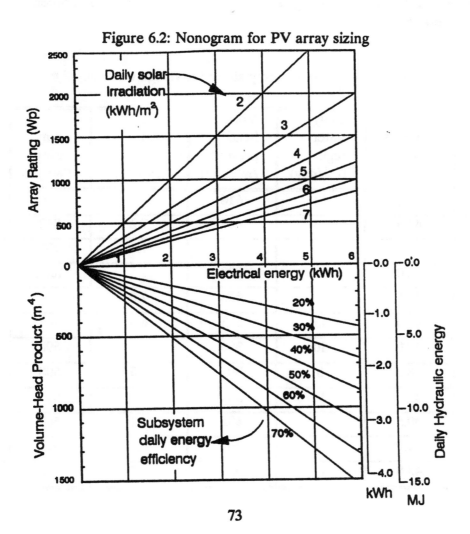

73

The nomogram in figure 6.2. relates volume-head product (hydraulic energy), available solar insolation and subsystem daily energy efficiency to give the necessary rated size of the PV array. Beginning in the lower half of the diagram, select the volume-head product for each month on the vertical axis. This axis has been labelled in both kWh/day and m⁴/day. Trace a horizontal line across to intersect the appropriate subsystem efficiency curve. From there, trace a line vertically to intersect the appropriate solar insolation curve (this should be defined for the tilted array surface angle and not for a horizontal surface). From that point trace a line left across to the horizontal axis, where the required array rating can be read in peak Watts (Wp). Copy this number into column 5 of the table as in the example in table 6.4.

Alternatively, you may use the equation:

$$W = 1000 \times VHP / (367 \times I \times e)$$

where W is the array rating in Wp, VHP is the volume-head product in m⁴, I is the daily insolation in kWh/m² and e is the subsystem daily energy efficiency expressed as a fraction.

The nomogram and the equation will give the same result. The choice between them is a matter of personal preference.

Table 6.4: Example array sizing for village water supply

Month	Total pumped head (m)	Daily water regu. (m³/day)	Volume-head product (m⁴/day)	Daily solar insol. (kWh/m²)	Array rating (Wp)	Actual volume pumped (m³/day)
Jan	44.0	14	616	6.9	810	17
Feb	44.5	15	668	7.2	842	18
Mar	45.0	17	765	7.4	939	18
Apr	46.0	18	828	7.5	1002	18
May	43.5	15	653	7.0	847	18
Jun	41.0	13	533	6.5	745	17
Jul	39.0	10	390	6.0	590	17
Aug	38.5	8	308	5.5	508	16
Sep	39.0	9	351	5.8	549	16
Oct	40.5	10	405	6.1	603	17
Nov	42.0	11	462	6.4	911	17
Dec	43.0	13	559	6.7	758	17

The design month is that with the largest array size needed to meet the conditions. In the example in table 6.4. this is April with an array size of about 1000 Wp. However, we must now check this against the maximum array size that was calculated from the well peak flow rate in section 6.2. If the design month array size is larger than this, then the maximum safe size must be used instead, or the well may be pumped dry. In the example, the maximum size was over 2500 Wp, and so it is not a constraint. We would therefore select our array size as 1000 Wp.

Motor/Pump Sizing - Although in practice the appropriate motor/pump system would be chosen by the system supplier, it is important for the potential buyer to have some comprehension of what is involved. This is also useful for calculating replacement costs in the life-cycle cost analysis in section 8. The motor and pump are usually one composite unit, and so are already mechanically matched.

The motor must be rated high enough to withstand the peak output of the array, and as motors are generally rated in terms of their maximum electrical input power, this must at least be equal to the array Wp rating. By arranging the PV modules in series and parallel, the array can be configured to match the motor voltage and current limitations.

Pipework Sizing - Once the array size to be used is known we can calculate the predicted peak flow rate required. This can be obtained from the peak hydraulic power produced and the head. The peak hydraulic power produced by the pump is given by the product of the peak array power output and peak subsystem efficiency. This can all be easily determined using the peak flow nomogram in figure 6.1. again. The array rating is given on the lower vertical axis, and for the appropriate peak subsystem efficiency the hydraulic power can be found on the horizontal axis. Using the upper half of the diagram with correct total head, the peak flow rate can be read off from the upper vertical axis. The dynamic head loss due to friction is then found using the graph in figure 6.3. First convert the flow rate from litres/sec to litres/minute by dividing by 60, and then choose a pipe diameter such that the dynamic head is very small compared to the total pumped head (i.e., at least less than 5% of total).

6.6. Estimating performance

It must be remembered that the sizing has been done for the worst case month, and does not represent the typical operating conditions. It is not absolutely necessary to calculate the water output in the remaining 12 months, because it is known that if the pump can meet the worst case, then it can cope all year round. However, it may be useful or interesting to know what the approximate output will be all year round, and this can be found in the following way:

The nomogram in figure 6.2. can be used in reverse, keeping the previously calculated array size constant, but using different daily solar insolation curves and total heads in the top half of the diagram for the different months. The hydraulic energy demand can then be read off in terms of volume-head product. Alternatively the array sizing equation given in section 6.5. can be rearranged and used thus (using the same symbols):

$$VHP = (367 \times W \times I \times e) / 1000$$

Both methods yield the same result for volume-head product, which should then be divided by the total head to give a daily delivery in m^3. Enter this as an extra column in table 6.4. for easy comparison with the ideal water requirements. This has been done for our example, using the array rating of 1000 Wp that was calculated in the last section, 6.5.

Figure 6.3: Frictional head loss as a function of flow rate

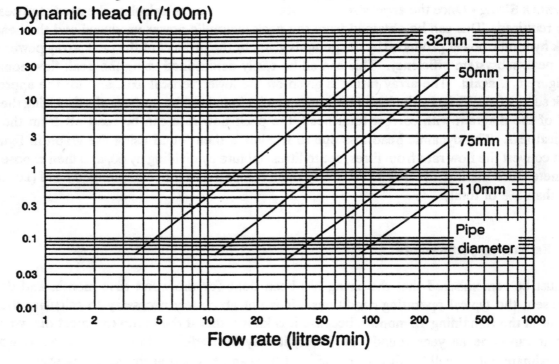

The following points follow from the discussion:

- If the well peak flow rate is not a limiting factor, then the array will have been sized for the worst month, and thus in each month at least as much or more than the required volume will be pumped.

- If the array size was limited by the well peak flow rate, then in some months (but not necessarily all) the actual amount pumped will be less than the specified required volume.

In the first case, if in most months the extra water pumped is very much larger than that required, then it may be wise to reassess the worst month requirements. If these can be reduced, then a smaller pump and PV array may be used which will, of course, cost less, while still meeting the original requirements for the rest of the year. On the other hand, if the larger array can be afforded, the extra water will almost certainly be put to some use.

In the second case a decision must be made as to what shortfall in water supply is tolerable. If the quantity pumped is much too small in almost all months, then a point will be reached when it may be decided that it is not worth investing in a solar pump at all. In this situation, more boreholes/wells will be needed whatever the pumping method, and with small quantities of water from each borehole handpumping may well be more appropriate.

In both situations, these factors are very much dictated by the relative costs (both capital and recurring) of the different pumping methods. A methodology for assessing these costs will be given in section 8.

It should be realised that the method used above will still be overly pessimistic in many situations, because it assumes that drawdown is always at its safe maximum. This is quite justifiable in that it adds a safety factor to our sizing calculations, but in reality the drawdown will be less than this at flow rates less than the safe maximum. Hence smaller array sizes may be required for given daily pumped volumes; conversely, for a given array size, greater daily volumes will be pumped.

Even if, for a particular well, the relationship between drawdown and pumping rate was exactly known, we would still face considerable difficulties: The pumping rate varies throughout the day with angle of the sun, and so the drawdown (and therefore head) will also vary accordingly. But pumping rate is itself dependent on the head. Thus this is an iterative problem in which pumping rate and head are not independent of each other, and also vary with time. This requires a methodology beyond the scope of this book, and best solved iteratively with a computer.

We can, however make some qualitative guesses as to how far we have departed from the likely reality. If the maximum safe array size is a strongly limiting factor in most months, then the drawdown will tend to be near the maximum most of the year and the daily volumes calculated will be reasonably accurate. However, drawdown is maximum only at full sun intensity (i.e., at mid-day) and so they will still be somewhat pessimistic.

If the maximum safe array size is far from being a limiting factor (e.g., if the design month array rating is 500 Wp and the maximum safe rating is 2500 Wp) then the drawdown is likely to be far less than the maximum for the required extraction rate. Thus a smaller array would be sufficient, or alternatively, more water could be pumped for the calculated array size. If the drawdown is a significant part of the total head then the over-sizing of the array could be quite large.

Clearly accurate array sizing is a matter of great complexity; the above methodology and discussion provide a means of determining a rough worst case array size and finding what limitations if any are likely to be imposed on the required delivery.

7. SITE AND SYSTEM EVALUATION FOR IRRIGATION PUMPING

The philosophy behind the design of a PV irrigation system is very different to that used for village water supply. Irrigation pumping is characterised by the need for very large quantities of water, and an uneven seasonal demand. Some crops require their maximum water supply for a relatively short period of their growing season, and all irrigation systems need to be sized for this peak. Therefore, pumps are effectively oversized for the rest of the year.

Due to the large daily volumes of water required, irrigation pumping is usually only practical from surface water sources, or high yielding shallow wells. Boreholes simply cannot yield enough water for more than micro-irrigation or watering of vegetable gardens, except in a few areas with exceptional ground water resources (e.g., Pakistan & Bangladesh). For example, a field of 1 hectare may have a water requirement of 75 m^3/day, equivalent in terms of village water supply to a village of 3000 people. Given that a surface water source is available, the water resource and drawdown considerations are not likely to be a constraint as they are in village water supply.

Instead the major constraint on the system is likely to be economic. For a PV pumping system to be worthwhile, the additional value of the crop that can be produced each year must outweigh the annual payments that the system incurs. Financial factors, such as the size of available loans from agricultural banks, must be taken into account and in-depth knowledge of local farming practices becomes essential. This obviously adds a great deal of complexity to the analysis, and the benefits can probably never be estimated with any great accuracy.

One of the greatest areas of uncertainty is in the estimation of crop water requirements, and the sensitivity of the crop to the water supply. This will vary greatly from crop to crop and with different soil types and rainfall regimes. While complex theoretical models exist for such calculations, they may not be consistent with the practicalities and economics of PV pumping, and would in any case be beyond the scope of this book. All that we can do is provide guidelines and a frame-work for further investigation.

Thus if there is access to a suitable water source we can say that the major factors that act as boundary conditions to the problem are:

(i)   The crop irrigation water demand
(ii)   The solar resource
(iii)   The economic benefits of irrigation

The first two are technical considerations, and with only these we could define a system of any size necessary (within the range of available systems) to irrigate the plot. However, it is the third factor that will limit the system size for most farmers, and compromises will almost certainly have to be made in the area of land to be irrigated, or in the quantity of water delivered in the peak months.

This methodology gives a basic means to determine the rough size of system and look at the basic technical feasibility of PV irrigation for a particular site. 'Rules of thumb' are used as guidelines to estimate the desired water demand and to calculate a corresponding system size. Points on irrigation economics are discussed, and with some local knowledge this can act as a starting point for a deeper analysis of the subject. Table 7.1. shows the method broken down into simple steps. These will be described in detail in the rest of this chapter.

Table 7.1: Steps in irrigation site evaluation methodology

I.	From a first look at the site determine the rough system layout, including size and shape of plot, type of crop, water source and likely distribution method.
II.	Estimate the irrigation requirements of the crop(s), the total area to be irrigated and the total desired water requirement on a monthly basis. Check that this does not exceed the available water resource.
III.	Gather monthly data on the availability of solar energy in terms of average $kWh/m^2/day$.
IV.	Calculate the array size that, for each month, would be necessary to provide the desired water supply, and initially choose the largest of these. Check the system for technical feasibility.
V.	Use the chosen array size to calculate the water that can be supplied each month.
VI.	Review the configuration and use economic considerations to optimise the system size and area irrigated.

7.1. Physical system layout

The first step is to take an overview of the site and determine the approximate system layout. The factors that will dictate this are (i) the position of the water source, (ii) the size and shape of the plot to be irrigated, and (iii) the application method. The water source may be surface water (i.e., pond, stream, canal, etc.) or a shallow well, and should not be more than a few metres below the surface. This will occur close to rivers and in drainage basins in permeable soil, where even well-water is effectively submerged surface water. The refill rate will be high and shortage of water should not be a problem. If water is only a metre or two below the surface a wide pit will provide a suitable water source, and can also act as a drinking point for livestock. Obviously, the closer the water source to the point of use the better. With a shallow water table it may well be worth the effort to dig another well rather than buy extra piping.

In areas where the water level is deep (i.e., > 10m) PV irrigation will seldom be practical. Yields of boreholes into rocky strata are typically low, and the power needed to lift sufficient quantities of water over such a high head will usually render the economics unviable.

The preferred system configuration for PV irrigation systems will almost always be a floating pump, given a surface water source or shallow well. These typically use a single stage centrifugal pump and are suitable for low heads and high flows. The main advantage is that they will move with changes in water level and can easily be lifted out and repositioned by a single person. Surface mounted suction pumps are also widespread, but can run dry and experience priming problems if left unattended.

The distribution system and the head required to drive it are the next points to consider. This is discussed more fully in section 3.5.3.1., but the two most practical solutions are a low-head drip system or hose and basin. Both are highly efficient (80-90%) and require a driving head of 1 to 2 metres. The main difference is in the cost. A low-head drip may cost $2500 per hectare while a hose and basin system may only cost a few hundred dollars. The method chosen will also depend on the crop being grown.

The slope of the field (if any) should be taken into account in calculating the necessary head to drive the distribution system. It is usual to use a header tank, often in combination with a low-head-loss filter, to give an even supply pressure. The height of this must be at least 1 to 2 metres above the highest point in the field (although most farmland in alluvial areas is flat). If you have any manufacturers data on the distribution system, ensure that pipe lengths do not exceed specifications. The required driving head may increase with the size of the network. A schematic diagram of a PV irrigation system is shown in figure 3.2.

It is essential at this point to draw a rough plan of the site and mark on the positions of the pump, main pipeline, and sub-mains for easy reference. The desired area (in hectares or m^2) of the plot to be irrigated should also now be decided upon. One hectare (ha) equals 10,000 m^2. Most farms in developing countries will be only 1 or 2 hectares, and in practice it may be unfeasible to irrigate more than this anyway. However, it is more likely that only the slightly more affluent farmers, owning perhaps 4 ha, will be able to afford (or raise the credit for) a PV pump. Another aspect of this is that very small farmers may have to use all of their land to grow staple crops, whereas the high cost of PV irrigation can only usually be justified if it is used on high-value crops. In any case, the area that we would wish to irrigate may have to be reconsidered at a later stage in the sizing.

From these site considerations we now have enough information to calculate the total pumped head. The total head can be said to be composed of three components: The height of the header tank, the depth to the water level and the dynamic head. If using surface water, drawdown is not likely to be a consideration, although more care should be taken in the case of wells. Even in good alluvium a relatively small system can induce a drawdown of 3 or 4 m in a high-yield open well (for a discussion of drawdown see section 6.2.). Similarly it should be possible to ignore the ynamic head: It will only be felt by the pump in the pipework up to the header tank, and this pipe can be oversized such that dynamic head is minimal. Therefore the total pumped head is just the sum of the header tank height and the water level depth, both of which should already be known from the above considerations. This should still be calculated for each month as there are likely to be seasonal changes in surface water levels or submerged shallow water tables. It is worth considerable effort to gain reliable data on this, as changes of only one or two metres are likely to be a significant fraction of the total head. For instance, failure to account for a 1 m drop in water level in an irrigation system whose total design head is only 4 m will result in a shortfall of 20% in terms of pumped volume. Equally, overestimating head by allowing high factors of safety can significantly increase the system cost.

As an example in the following sections we will take a level field of 1.5 ha, with a header tank positioned at a height of 2m above the surface. We will assume the same climatic conditions as for the village water supply example in the previous chapter. The pump is a floating unit in a water-hole whose depth to the water level (in m) varies annually as follows:

Jan: 2.7	Feb: 3.0	Mar: 3.5	Apr: 4.1	May: 3.2	Jun: 2.5
Jul: 2.2	Aug: 1.9	Sep: 2.1	Oct: 2.2	Nov: 2.4	Dec: 2.5

The monthly total heads are then just these figures plus the header tank height of 2.0 m.

7.2. Crop water requirements

To find the daily volume of water that the pump must produce, it is necessary to find an estimate for the crop water requirement. This is most conveniently defined in litres per m^2 per day. This is perhaps the most difficult part of the evaluation, as the crop water requirement depends on many factors, none of which can be known with any great accuracy.

Firstly it will vary between different crops, and its annual distribution will of course depend on the cropping pattern. At some points of its growing season a crop will need more water than at others. This is most likely to be when it is growing at its fastest. In addition, crops with a larger canopy (i.e., more or larger leaves) will loose water through them more quickly by evaporation. This is called evapotranspiration. The type and condition of the soil will also affect its ability to hold water in the root zone.

All this must be balanced against the rainfall that can be expected in each month, with the irrigated requirement making up the shortfall. In many tropical areas (e.g., Monsoon regions) the rainfall distribution can be predicted quite reliably, and data will be available from the meteorological office, universities or agricultural institutions. A farmer will know from experience in which months to expect rain and when little irrigation will be needed.

To make most effective use of the solar pump it will probably be advisable (and is usually necessary anyway) to grow more than one crop per year. Then the demand on the pump is more even all year round. For instance a cereal crop could be alternated with a vegetable crop. It should be clear that all these factors require in-depth local knowledge, and that generalisations cannot be made with any accuracy.

However, rules of thumb have been developed from traditional irrigation experience, some of which we can apply to PV pumping to get a feel for the quantities involved. One of the most useful rules of this type has come out of work on drip systems, and is given as follows:

- In hot dry climates use 7-8 litres/day per m^2 of crop canopy.
- In cooler or more humid places 5-6 litres/day per m^2 of canopy is sufficient.

This can be converted to an equivalent in m^3/day per ha of crop canopy, which is more convenient to use in our sizing calculations. Simply multiply the original figure by 10: For instance, 8 l/day/m^2 equals 80 m^3/day/ha. So if a crop at a certain stage in its growth covers 50% of the ground area, then a one hectare plot will have a total canopy of 0.5 ha. Therefore (assuming the hot dry regime) the volume required per day would be 80 x 0.5 = 40 m^3 for each hectare of crop. Clearly, the fraction of canopy cover will increase as the crop grows, and will depend on how closely the crop has been planted. However, this is something that can be estimated with only a little local knowledge of the crop and the way it is planted, and does not need detailed technical information or calculations.

To find the actual daily pumped volume required the volumetric efficiency of the application system must also be taken into account. This is done by taking the total field water requirement found above, and dividing it by the application system efficiency expressed as a fraction. For instance, for an efficiency of 80% divide by 0.8. The efficiencies for different application systems are given in table 3.3.

The determination of the daily pumped water requirement described above can also be performed graphically using the nomogram in figure 7.1., instead of by calculation. The methods yield the same result and which one to use is a matter of personal preference. To use the nomogram, start on the left-hand horizontal axis labelled 'size of plot' and select the appropriate number of hectares to be irrigated. Trace a line vertically up to the appropriate canopy fraction line and then right to the 'canopy area' axis. Extend this line further across to the relevant water requirement line (in l/day/m^2 canopy) and then trace a line downwards through the 'volume at field' axis to the appropriate volumetric efficiency line. Finally, trace a line left to the lower vertical axis to read off the necessary pumped volume in m^3/day. Thus we proceed in a clockwise direction around the nomogram. Alternatively it can be used in reverse (anti-clockwise) to find the size in hectares of irrigable land given a certain pumped volume, volumetric efficiency, crop water requirement and canopy fraction.

It should be noted that the nomenclature used in this book may not be universal. For instance, you will often find the crop water demand called the 'reference crop evapotranspiration' and expressed in terms of mm/day. This is an equivalent water depth, and is equal to litres/day/m^2. That is to say, that a litre of water distributed over a square metre would be 1 mm deep. Similarly, with more advanced methods it is usual to use a 'crop growth coefficient', which is defined slightly differently to the simple canopy cover fraction that we have used, but can be taken as the same figure for most practical purposes.

Figure 7.1: Water demand nomogram for irrigation

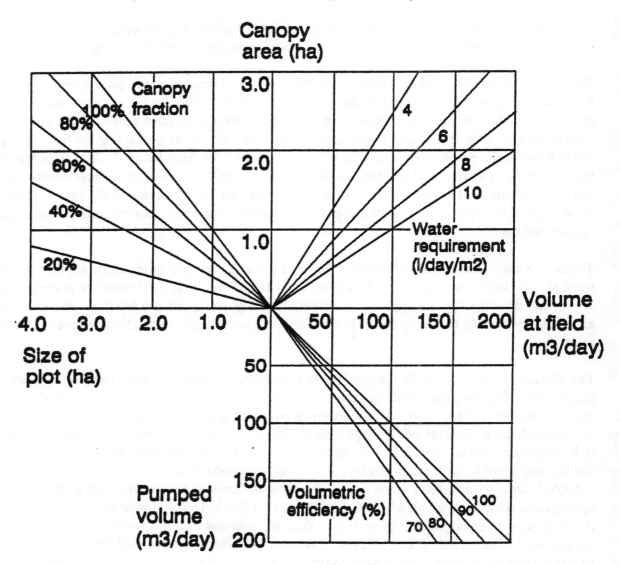

The method that has been described in this chapter is suitable as an initial rough guide and is demonstrated below on our example from the previous sub-section. Assume then, that our example 1.5 ha plot is in a very hot, dry region, and that two crops per year will be sown: An estimate of the canopy cover fraction, and the corresponding total pumped volume per day might look like that in table 7.2. We will assume 8 l/day/m² canopy (i.e., 80 m³/day/ha canopy) in the hot season (Dec to Jun) and 6 l/day/m² (60 m³/day/ha) in the cool season (Jul to Nov). Using a low-head drip system we will assume an application efficiency of 0.85.

In the example in table 7.2., Column 3, the total canopy area is simply formed by multiplying the crop canopy fraction by the size of the plot (1.5 ha). Column 5, the daily volume needed on the field, is formed by multiplying the total canopy area (ha) by the crop demand (l/day/ha). Column 6, the total pumped volume is the daily volume needed on the field (m³/day) divided by the application efficiency (0.85). It is the total pumped volume that we will use in the next section on array sizing.

Table 7.2: Example calculation of monthly irrigation requirements

Month	Canopy fraction	Total canopy ha	Crop demand m³/day/ha	Daily vol on field m³/day	Total pumped vol m³/day
Jan	0.1	0.15	80	12	14
Feb	0.2	0.30	80	24	28
Mar	0.5	0.75	80	60	71
Apr	0.7	1.05	80	84	99
May	0.8	1.20	80	96	113
Jun	0.0	0	0	0	0
Jul	0.1	0.15	60	9	11
Aug	0.2	0.30	60	18	21
Sep	0.3	0.45	60	27	32
Oct	0.4	0.60	60	36	42
Nov	0.5	0.75	60	45	53
Dec	0.5	0.75	80	60	71

However, before proceeding further it is necessary to consider whether the water source is likely to be able to sustain the extraction rates we have calculated. If in just one or two months the water demand is impractically high, then it may be possible to reduce it without significant damage to the crop. If this approach is not possible, then the area to be irrigated must be reduced.

In order to do a complete and professional evaluation more details are required. One such method calculates a figure for crop evapotranspiration based on a reference evapotranspiration and a crop 'growth stage coefficient' for each month. Guidance on these items is available in the 1977 FAO publication 'Crop Water Requirements' by Doorenbos and Pruitt. This is directed at conventional irrigation but contains much useful general information that can be of use to the potential solar PV pump user.

7.3. Array sizing

Before we can proceed with sizing the array and pumpset, we need to determine the availability of the solar resource for the site. This should be recorded for each month in $kWh/m^2/day$ as is described in section 6.3. In the example the same data has been used as for the village water supply example in the previous chapter.

We should now have enough information to carry out the system sizing. Having gathered together monthly data for (i) the total head from section 7.1., (ii) the desired pumped water requirement from section 7.2., and (iii) the solar resource, construct a table like that shown in table 7.3., and enter the figures in columns 2, 3 and 5 respectively. The figures used in table 7.3. are for our example situation.

In the fourth column we must calculate the daily volume-head product (in m^4) for each month as this is proportional to the hydraulic energy required. For each month, simply multiply together the total pumped head from column 2 and the daily water requirement from column 3, and enter this in column 4.

A final item needed before we can calculate the array size is the daily energy efficiency of the motor-pump subsystem. In the discussion of physical system layout in section 7.1., the system motor-pump configuration should have been decided upon, and typical sub-system efficiencies are shown in figure 3.7. A floating pumpset will have an efficiency of 25 to 35% and a surface suction pumpset perhaps nearer 25%. A figure of 30% will be taken for our example case which we have defined as a floating unit.

The size of PV array to be used can be deduced from the volume head product, the daily solar insolation and the daily energy efficiency. This can be done either by use of the nomogram in figure 6.2. or by the equation given in section 6.5. Both methods yield the same result and are described fully in section 6.5. The results (in Wp) should be entered in column 6 of table 7.3. If, as in the example, the daily volume-head product varies very much more month-by-month than the solar insolation, it is only necessary to calculate the array rating for the larger values of volume-head product. However, in the example, all have been calculated for the purpose of illustration. The design month is that with the largest array size needed to meet the conditions. We then know that the water requirements in all the other months can be met. In the example in table 7.3. this is May, with a rating of about 760 Wp.

86

Table 7.3: Example array sizing for PV irrigation

Month	Total pumped head (m)	Daily water requ. (m³/day)	Volume-head product (m⁴/day)	Daily solar insol. (kWh/m²)	Array rating (Wp)	Actual volume pumpable (m³/day)
Jan	4.7	14	66	6.9	87	123
Feb	5.0	28	140	7.2	177	120
Mar	5.5	71	391	7.4	480	113
Apr	6.1	99	601	7.5	728	103
May	5.2	113	588	7.0	763	113
June	4.5	0	0	6.5	0	121
Jul	4.2	11	46	6.0	70	120
Aug	3.9	21	82	5.5	135	118
Sep	4.1	32	131	5.8	205	118
Oct	4.2	42	176	6.1	262	122
Nov	4.4	53	233	6.4	331	122
Dec	4.5	71	320	6.7	434	125

At this point it is necessary to review the figures that have been calculated for their technical feasibility. A typical small irrigation system might have an array rated at 500 Wp, although PV pumping systems of up to 1000 Wp are not uncommon. Certainly it is rare to find systems over 2000 Wp and they would be unaffordable for the small farmer. Therefore, review the chosen PV array rating and if it seems unfeasibly large then choose a lower rating. This means that either the crop will receive less water than originally desired in some months, or the area to be irrigated must be reduced. It is worth browsing through the list of available systems in appendix D to see if systems exist that will fit your criteria.

To size the motor/pump unit and the pipework (as far as the header tank) the procedures described in section 6.5. for the village water supply case may be used.

7.4. Performance and optimisation

There is great scope in PV irrigation system design for variation and optimisation. Fortunately, the typical farmer in developing countries is well used to such improvisation to make best use of what little he has, and will, within reason, be able to adapt to changing situations. In the paragraphs below we will examine some of the factors and relationships that may affect the decision making process once a 'first draft' design has been completed.

Having chosen an array rating it is now possible to calculate the actual quantity of water that a system of that size could produce per day given the solar insolation for each month (and assuming enough water available). This can be done using the nomogram in figure 6.2. in reverse, or using the equation, both as described in section 6.6. in the previous chapter. These figures can be entered in an extra column in the table, as has been done for the example figures in column 7 of table 7.3.

If the design month array size has been used then there will be no problem as to the quantity of water delivered to the crop. By definition, if the system can cope in the design month then it can cope in all others. However, this means that the PV array will probably be oversized for the rest of the year. If costs are critical and one month has a much greater required array rating than most others, then it may be possible to reduce the irrigation quantity for that particular month, and thus reduce the design month array rating. Of course, this can be dangerous if it is not known how the crop will respond to this. Agricultural knowledge is essential with this sort of speculation, as the supply in the peak demand month may be critical to the success of the crop.

Another solution to the problem of over-sizing may be to consider new cropping patterns to spread the demand more evenly. Once the actual daily quantities of water have been calculated it should be relatively straight-forward to find, perhaps, another crop that could make use of the possible excess supply during some periods of the year. Farmers will usually have no problem in finding uses for extra water, but utilising that water to their best advantage will take some thought and planning.

It is also a reasonably common practice to reduce the necessary array size by the use of movable arrays. These would typically be mounted on posts such that they can be rotated in two planes. In the peak demand month(s) they could be moved 3 or 4 times a day by hand, and thus always remaining facing the sun. This will dramatically increase the daily water output compared with using a fixed system, and will therefore be of great importance to the economics of the system. Changing the orientation is not a problem as far as labour is concerned, as rural irrigation systems will not be left unattended for long periods, and re-alignment would probably only be necessary in the peak demand months. By boosting the output in this way, a smaller array can be used, and the over-sizing is not so severe for the rest of the year. Calculation of the output that can be expected using this technique is not straight-forward, and should be carried out by a PV system manufacturer or supplier. However, this is certainly an area that should be looked into before choosing a system.

If it has been necessary to use a smaller PV array than that found in the above method, then there will be a shortfall in the originally specified daily quantities. As stated above, this will mean either accepting a lower water supply per m^2 or reducing the area irrigated. If the shortfall is very significant throughout most of the year, a decision must be taken as to whether it is worth using solar pumping at all. As always local knowledge is essential is discerning what impact a shortfall of water at a certain point in a crop's development will have on the final yield.

Ultimately, it is the economics of the situation that decide whether or not to use PV pumping, and to what degree. Clearly, if the cost to the farmer is greater than the extra return he receives from his crops due to irrigation then there is no point in pursuing it. Exceptions to this may be where irrigation is used on a staple crop to provide security of food supply in areas where the soil or climate make the harvest unreliable. However, it is more normal to use PV irrigation on high-return cash crops which could not normally be grown, with perhaps a portion of the water used to provide security of staple crops if necessary.

To the small farmer financial considerations are everything. Even if given a grant towards the capital cost of an irrigation system a farmer in a developing country will have to borrow the remainder from a bank or other lending institution. This will of course mean annual repayments, the size of which will depend local financial factors. He will also most likely have to pay for annual maintenance and running costs himself. Thus in most cases life-cycle costing as dealt with in chapter 8 is not relevant from the farmers point of view, and it is the annual payments (loan repayments and maintenance) that he must balance against the return he receives from the crop. However, chapter 8 can be used to estimate both capital costs and recurring costs of PV pumping systems. Thus some iteration between this chapter and chapter 8 may be necessary as different scenarios are tried and costed. As the loan is paid off, the annual payments will decrease, and so initially a lower level of profitability will be acceptable. For most farmers in the developing world it is unlikely that PV pumping could ever be economically viable without some form of grant or other financial aid, at least for the foreseeable future.

One way to cut costs is by making economies in the distribution system, although this does not necessarily mean making great sacrifices in terms of efficiency. For example, a low-head drip system may cost in the region of $2500 per hectare, which is likely to be a considerable portion of the total system capital cost. It may be advisable to begin with a hose and basin system, costing only a few hundred dollars, and perhaps convert to a drip system at a later date. This allows the farmer to purchase as big a pump as he can afford, and then expand his distribution system as his finances permit.

Thus, within the bounds of what is affordable, a farmer should ideally optimise his system size on economic grounds. This adds a whole new level of complexity and uncertainty. Suppose that a first draft analysis for a certain irrigated area, water demand and cropping pattern is costed and found to be uneconomic in terms of the returns from the crop versus the system cost to the farmer. It may be that a larger system would benefit his crop to such an extent that it balances his slightly increased payments on the PV system. On the other hand, perhaps this would add to the system cost without providing much extra crop income, and that what is actually needed is a smaller system to make the economics balance.

If PV pumping is at all viable for a situation, then in general there will be a size range over which the PV system provides a net benefit. This relationship can be illustrated schematically by the graph in figure 7.2. The horizontal axis represents increasing water supply per m^2 to a particular crop. The curved line represents the additional income generated from an irrigated crop. This starts at zero and rises steeply to a plateau where additional water provides little extra benefit. The straight line represents the annual PV system payments the farmer must make. These start above zero, but rise approximately linearly as system size increases. The shaded area between the two lines (i.e., when the PV costs are lower than the additional crop return) is the range in which PV irrigation is profitable. The point in this range at which the two lines are furthest apart represents the greatest benefit and therefore represents the optimum system size. Of course, it is quite possible that for a given situation the curves will never cross and the PV curve is always higher than the crop return curve: In this circumstance PV irrigation can never be of net benefit.

Diagrams such as that in figure 7.2. serve to illustrate in a simple way the relationships involved in irrigation economics, but in practice could never be accurately constructed, as there are too many unknown quantities and uncertainties. To even roughly determine the optimum system size requires at least some knowledge of local climate, soil, farming practices, crop growth cycles and financial conditions. The best source of knowledge must be the experience of others, and similar projects should be sought out and their lessons learned.

Figure 7.2: Variation of irrigation viability with supply rate

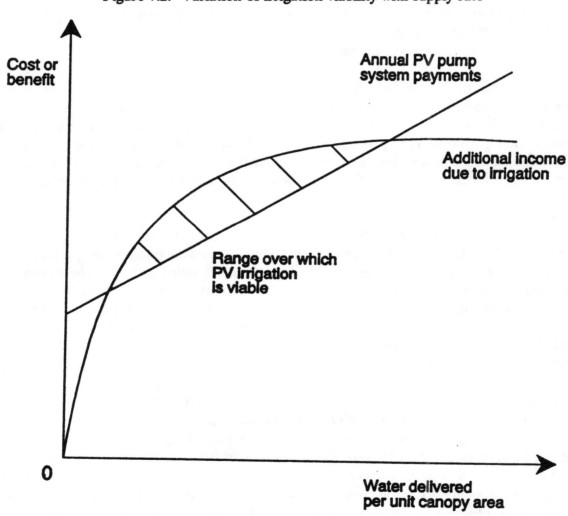

8. PRACTICAL COST APPRAISAL

8.1. Economic evaluation

Installation and operation of any pumping system requires a long-term financial commitment and the consequences of inadequate assessment beforehand can be dire, especially for poorer communities. It is therefore important to discuss the way that various factors may affect the economic or financial viability of an installation.

In many situations, the potential pump users will be constrained by hydrological factors to the sort of pumping system they can use, and as such do not have a real choice. In the majority of situations in which hand pumps are presently used, the wells may not yield enough water for the installation of a mechanised pump. Extraction of more water would either quickly pump the well dry, or may increase the drawdown to such a degree that PV pumping is not feasible anyway. The potential well yield is thus the primary limiting factor on all forms of mechanised pumping.

Clearly for renewables, the next most important factor is the level of the resource. In the case of PV, the size of the array required is inversely proportional to the insolation in $kWh/m^2/day$, and so is related in a straight-forward way to the system cost. However, resource assessment for windpumping can be far more critical. This is because the power obtainable for a given rotor size is proportional to the cube of the velocity. Hence a reported mean wind of 4 m/s instead of 3 m/s would mean an under-sizing of the system by a factor of nearly 2.5. This would obviously have a very serious effect on the unit cost of the water produced.

Diesel engines are usually oversized and fairly cheap to buy, and so accurate sizing is not so critical as for capital intensive systems like solar and wind. However, the two factors that are of more importance are availability and cost of fuel and maintenance services. Fuel prices can vary enormously between different countries, and also between different areas of the same country. Some governments subsidise diesel fuel, while other tax it heavily, and this is where the distinction between economic and financial viability becomes important. The international market price of diesel which is fairly constant, and will typically be lower than the price to the user, which will include taxes, transportation costs and retailer's margins. Hence the situation may arise where diesel may be the cheaper option on an economic basis, but may not be so from the point of view of the user. In some rural areas, pumping time lost due to non-availability of fuel or waiting for maintenance can also be a factor that can affect the economics.

Taxes and subsidies are a very complex area, but can make a crucial difference to the financial viability. For instance some countries, such as India, will give tax relief on any renewable energy equipment. Import taxes and regulations may have some effect on the cost of high-tech equipment that must be imported (i.e., PV modules).

An important factor to consider, particularly when comparing PV pumping to hand pumping is the cost of the borehole. Drilling is very expensive in remote locations and the borehole cost may represent a significant portion of the total capital cost. As the borehole cost will be the same whether a hand, diesel or solar pump is used, the quantity of water that can be pumped from one borehole will affect the unit water cost of the system. For instance, a handpump and a PV pump may each require a borehole costing $10,000, but the handpump may serve only 200 people, whereas the PV pump could serve 1000 given sufficient well capacity.

8.1.1. General methodology

Due to its high reliability and 'stand-alone' operation, solar pumping would appear to be the ideal choice for remote rural areas. However, the potential buyers in these areas are also likely to belong to the poorer communities. Therefore, although solar pumping may be technically viable, it is vital to make a thorough examination of the costs involved, and compare them with the alternatives. This is done in general economic terms for village water supply in appendix I, and more specifically (for just PV) in appendix H. However, each real scenario is different, and a practical cost appraisal should be carried out individually for each site using the appropriate details.

It is important at this point to recognise that the method of cost evaluation should change depending on who is doing the appraisal:

A financial appraisal is an evaluation from the purchaser's point of view. In this case taxes, subsidies and the effect of spreading the capital cost over several years by means of a loan are all taken into account. This is of course far more difficult to perform without some local knowledge, but is all-important from the user's point of view. For most development projects (either village water supply or irrigation) much of the capital cost will need to be provided in the form of aid grants or other financial assistance. However, the remainder of the capital must be paid for by the community (or the farmer) and they will almost certainly have to take out a loan to fund this from an agricultural bank or other lending institution. Clearly, each year they will need to make repayments to this loan, the size of which will depend on the size of the loan and the local financial conditions. It is very likely that the village/farmer will also have pay for annual maintenance and operating costs of the system. Therefore, the crucial factor to the user is not the lifetime economics of the system, but the cash flow from year to year. The community must be able to afford the sum of the annual repayments (if any) and the annual running costs. In the case of irrigation, the additional income from the extra crop grown must be sufficient to cover these items. As the loan is paid off, the annual costs can be expected to decrease, although to maximise affordability it may be advisable to spread costs over as long a period as possible.

The economic approach looks at viability from a governmental viewpoint, seeking to compare the value to the economy as a whole, over a long timescale. When comparing solar pumping with, wind, animal, hand or diesel powered pumping, we need to be able to calculate the real cost to the community of each system. It is not sufficient simply to compare the initial cost of the system, as this takes no account of the costs or benefits from the system in the future. Equally, costs and benefits must be used that are free from taxes, subsidies, interest payments, etc. (e.g., the international price of diesel fuel, with some allowance for distribution costs, would be used rather than the local price).

Both financial and economic appraisal are equally important in their own rights, and should be used according to the emphasis that is required. In the subsequent section we will detail the method for an economic assessment, as this is the most general case. Methods for calculating the components of recurring cost (such as maintenance, spare parts, diesel fuel, labour costs) required for the financial evaluation will be covered in various sections of the economic methodology, but should be used at their present value.

There are several ways that you may come across to assess the cost competitiveness of different pumping systems:

Pay-Back Period: The length of time required for the initial investment to be repaid by the benefits gained.

Rate of Return: The benefits gained are expressed as a rate of return on the initial investment.

However, the above two methods have two disadvantages. Firstly, it is not always easy or meaningful to express the benefits gained in monetary terms, and secondly, they do not take full account of the longevity of the system, or any future costs associated with the system over its lifetime.

The most complete approach, which we will deal with in detail in this chapter, is to calculate the Life Cycle Cost. By this method, not just the initial costs, but all future costs for the entire operational lifetime of the pumping system are considered. The period for the analysis must be the lifetime of the longest lived system being compared.

For instance, a solar pump costs more to buy than a diesel pump, but the modules (which usually make up most of the system cost) will last for about 20 (or more) years. The diesel pump might last 10 years, and use a certain amount of fuel each year. In this case the analysis period is 20 years, and the cost of a replacement diesel plus 20 years worth of fuel must also be included for the diesel option. In addition, the costs of maintenance and repair for the two systems over their whole lifetime must be included. Depending on the exact figures, either the solar pump or the diesel may work out cheaper overall. It should also be pointed out that a life-cycle analysis can also be done financially if appropriate, using local prices, taxes, subsidies, etc.

For each pumping system on which we are going to perform a life-cycle cost analysis, we need to identify all the initial and future costs. These can be generally divided into the following six categories, each of which is discussed in more detail in the next section.

- Initial capital cost
- Installation
- Operation and maintenance over whole lifetime.
- Fuel (only for diesels) over whole lifetime.
- Replacement of components during lifetime.
- Additional income from extra crops in irrigation case.

Because the value of money changes with time it would be unrealistic to simply add up the future costs as they stand. Future costs and benefits must be "discounted" to their equivalent value in today's economy, called their 'present worth' or PW. To do this, each future cost is multiplied by a 'discount factor' calculated from the 'discount rate'. A discount rate of 10 % per year would mean that in real terms it makes no difference to a farmer whether he has $100 now or $110 dollars in one year's time. Conversely, a cost of $110 dollars one year from now has a 'present worth' of $100. This can be a confusing concept, and it should be stressed that the change in the value of money expressed by the discount rate is not the change due to general inflation. It reflects the return the farmer could have got on his money had he invested it. For this reason it is sometimes known as the 'Opportunity cost of capital'. This is typically 8-12 % in most economies. High discount rates mean that a low value is placed on future costs and benefits, and a high value is placed on present capital. The discount factor is calculated from the inflation rate and the discount rate.

As it is usual to work in real terms, the commodity-specific inflation rates and discount rates used should be defined as relative to the general inflation rate. Therefore inflation is taken as zero when calculating the discount factor, unless the price of a commodity (i.e., diesel fuel) is expected to rise faster or slower than general inflation.

All the components of the total life-cycle cost in the above list should be expressed in terms of their present worth before summing them, and an example of such a calculation is shown in the next section (8.1.2.).

8.1.2. Calculation of present worth

There are two types of calculation that are commonly used in life-cycle costing when expressing a future cost or benefit as its present worth. The first is used to calculate the present worth of a single payment, say the replacement of a motor/pumpset after ten years. The second is used to calculate the total present net worth of a recurring cost, such as annual fuel or maintenance costs. This is the sum of many discounted single payments over the analysis period. To simplify these calculations and obviate the need for complex equations the relevant PW can be found by multiplying the actual cost by a factor that can be found from tables in appendix F. The formulae from which the tables were calculated are also given at appendix F. Note that the inflation and discount rates are expressed as fractions rather than percentages, (e.g., 23 % is 0.23).

(a) For a single future payment:

For a single payment Ps ($) to be made in the future, the present worth, PW, is found by multiplying the payment Ps by a factor Fs, found from table F.1 in appendix F. To find Fs from the table you need to know the discount rate, the commodity-specific interest rate (relative to general inflation), and the number of years in the future that the payment is to be made.

$$PW = Ps \times Fs$$

Example: It is estimated that a new motor/pumpset will be required for a certain solar pumping system in ten years time. We will assume that a new pumpset presently costs $1000, that pumpset prices do not change relative to the general inflation rate, and that the discount rate is 10 %. Using table F.1. of appendix F. with $i = 0.0$, $d = 0.1$ and $n = 10$ (the number of years hence that the payment is to be made), this gives a discount factor of 0.39. Therefore the present worth of this future cost is:

$$PW = \$1000 \times 0.39 = \$390$$

(b) For a recurring annual payment:

For a payment of Pa occurring annually over a number of years, the net present worth is found by multiplying the annual payment Pa by a factor Fa, found from table F.2 in appendix F. To find Fa from the table you need to know the discount rate, the commodity-specific interest rate (relative to the general inflation rate) and the number of years for which the payment is to be be made.

$$PW = Pa \times Fa$$

Example: The fuel costs for a particular diesel pumping system are $50 per year, and in a rising oil price scenario, we might assume that diesel fuel prices will rise at 5 % above the general rate of inflation. We will also assume a discount rate of 10 %. Using table F.2. of appendix F., with $i = 0.05$, $d = 0.1$, and the number of years being the length of the analysis period, say 20 years, we get a cumulative discount factor, Fa, of 12.72. So the present net worth of the diesel fuel costs over the whole analysis period is:

$$PW = \$50 \times 12.72 = \$636$$

8.2. Cost appraisal of water pumping

8.2.1. General procedure

As discussed in section 8.1.1., to perform a life-cycle cost comparison we must sum the contributions from the items in the following list:

- Initial capital cost
- Installation
- Operation and maintenance over whole lifetime.
- Fuel (only for diesels) over whole lifetime.
- Replacement of components during lifetime.

However, to find all these components involves the use of many different sources and types of data, and so a strictly organised approach is needed. In figure 8.1., the the whole sizing and costing procedure has been divided up into logical self contained steps. This step by step approach is based on a life-cycle costing of the whole system. It takes into account each of the identifiable costs, but, for the sake of generality, ignores the benefits gained by the users. These are often difficult to quantify, and will be very dependent on local conditions. For example, it does not tell us if the return from the extra crops produced by irrigation offsets the extra cost of the water, as detailed knowledge of the particular crop and its characteristics are required. In the case of village water supply the benefits are more qualitative, such as improved health, less drudgery and lower infant mortality rates.

In short, life-cycle costing will reveal which is the most viable pumping system among various options to deliver a given water demand, but will not tell us if the pumping system is viable in an absolute sense (i.e., if the benefits gained are of greater value than the cost of the system). The final result of the analysis can be expressed more meaningfully as a 'unit water cost' in cost per m^3 of water. Of course, the least cost solution may not be the final choice, since other non-economic factors should be taken into account. Reliability, user acceptance and spares or fuel availability must all be considered. However, a cost appraisal is a necessary step in the decision making process.

8.2.2. Step-by-step procedure for life-cycle costing

In the following few pages, we will examine the steps necessary to perform a full life-cycle costing of solar, wind, diesel and hand powered pumping systems, roughly according to figure 8.1. This process has been divided into 4 simple steps.

The components of irrigation and village water supply pumping systems (for the purposes of a cost appraisal) are listed in table 8.1. This applies to pumping systems generally and is not just restricted to solar powered systems. The entries in brackets are optional and may not be included in every system.

96

Table 8.1: Components of pumping systems

Irrigation System	Village Supply System
Water Source	Water Source
Power Source	Power Source
Pump	Pump
(Storage Tank)	Storage Tank
Water Conveyance Network	(Piped Distribution)
Field Application Method	

It is assumed at this point that by using the sizing methodologies for either villarge water supply (chapter 6) or irrigation (chapter 7) that an appropriately sized PV system has been defined. The hydrological data concerning daily demand, total head and volume-head product that were calculated as part of the PV sizing apply equally well to the other options.

General comments and quick reference data for the sizing and economics of wind, diesel and hand pumping are contained in a concise form in appendices G.1, G,2 and G,3.

If during the costing process it becomes apparent that the system is far beyond what is viable in monetary terms, then it may be necessary to go back to chapters 6 or 7 and review your site evaluation.

The first step should have already been completed for the PV case and relates to sizing the system and calculating the energy requirements such that the costs can be found.

The data required in step one are:

- Hydrological data: Daily demand, total head, volume-head product, and well drawdown data.
- Meteorological data for the site (Sun and wind)

If you are not making a comparison with diesel, wind or hand pumping, but simply costing a PV system then go straight on to step 2.

Step 1: Size the power source and pump

The 'design month' is the month for which the largest size of power source is needed. This will depend on the water demand, but for the renewables it also depends on the resource availability. This is because lower sunlight levels (or wind speeds) mean that more PV modules (or a larger wind turbine) are necessary to produce the same power output. Thus we must calculate the necessary system size for PV, wind, diesel and hand pumping on a monthly basis, and then select the largest size for each option. We then know that the system will be adequate for all other times of year.

For solar pumping the design month will be the one that needs the largest array rating. Section 6.5 (or 7.3 for irrigation) explains how to estimate this figure. The main result that we need for economic purposes is the array size in Wp, and the capital cost of the rest of the pumping system will tend to scale with this. Solar pumping systems are usually bought complete, and it is not necessary to know the exact motor or pump specifications when ordering. However, for a life cycle cost analysis the size of the motor/pump unit is needed (sectiom 6.5) to give some indication of its replacement cost.

Figure 8.1: Steps in a life-cycle cost analysis

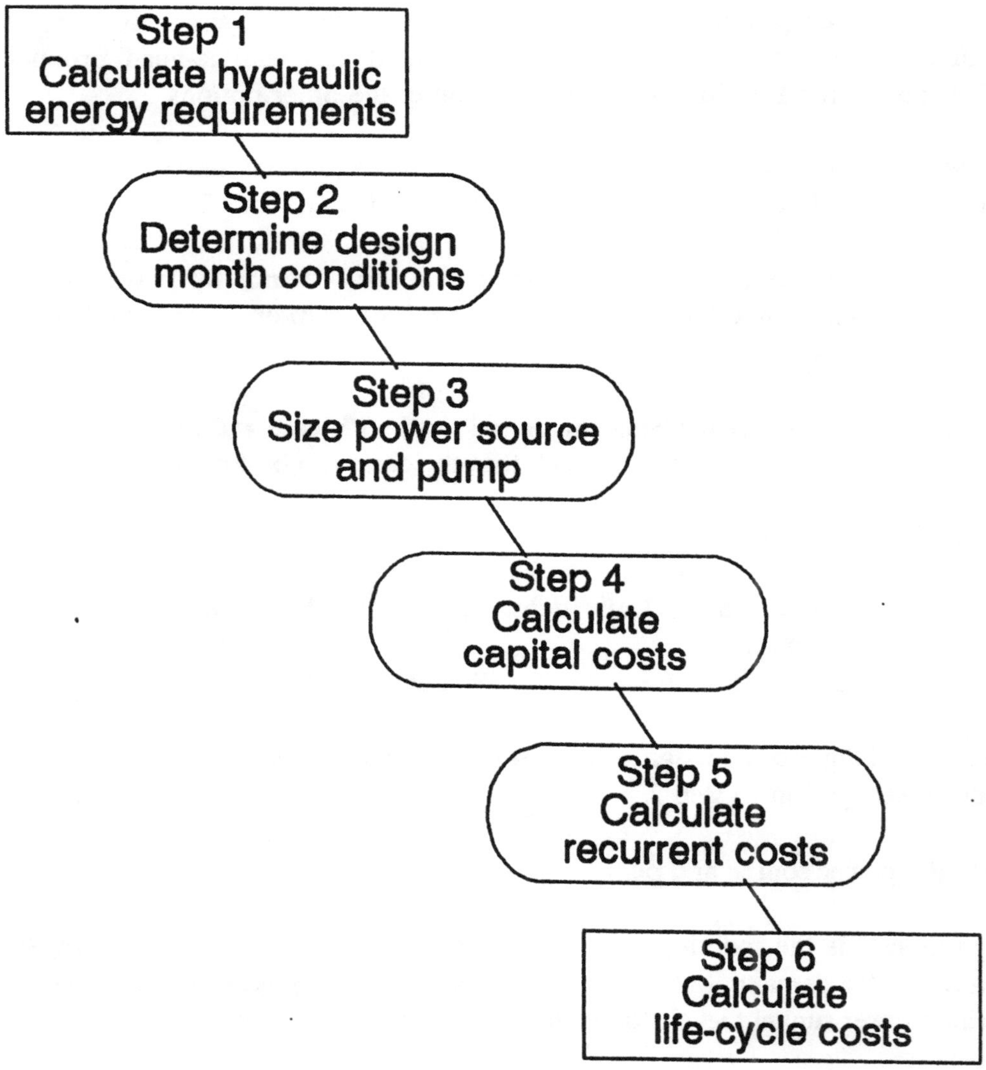

For wind pumping, the design month is that which requires the largest rotor area. We cannot simply use the ratio of demand/windspeed because the power extracted by the rotor is non-linear and depends on (windspeed)3. So we must find the rotor size required for each month, and use the month needing the largest sized rotor for windpump size. Figure G.1 in appendix G shows a graph that enables this quantity to be read from the vertical axis if the mean windspeed and volume-head product are known. Although windpump prices tend to be proportional to rotor area, windpumps are usually specified by their rotor diameter, and so for an area A the diameter, d, is given by:

$$d = 2 \times SquRoot(A/pi)$$

Alternatively, the rotor diameter can be found from the left hand half of the nomogram G.1. From the rotor area on the vertical axis, trace a horizontal line (left) across to meet the curve, and then trace a vertical line downwards to read off the rotor diameter on the horizontal axis.

The design month for diesel and hand pumps is simply that with the highest water demand. It should be noted that the design month for solar, wind and diesel/hand power may well be different, and data for the appropriate months should be used for the different power sources in subsequent steps. For the conventional option, unless water requirements are extremely large, the smallest size of diesel pump should be sufficient (ie. up to a volume-head product of 6000 m4 per 12 hours of operation). A small kerosene engine may provide up to half of this amount.

In the case of hand pumping, the 'system size' is the number of people required to perform a given hydraulic duty. Appendix G.3 gives a graph (figure G.3) from which we may calculate this number for the given head and design month daily water requirement.

However, handpumps are limited to what a person can comfortably use, and so will be of a standard size. At a certain number of people, the pump will be in operation almost continuously, in which case more than one pump may be required. The limit for one pump is around a volume-head product of 250 m4/day. For village water supply, a figure of one pump to supply 250 people is often used.　.

For the next three steps the following data are required:

Economic:　　Period of analysis
　　　　　　　Discount rate
　　　　　　　Inflation rates

Costs for each component:
　　　　　　　Capital costs
　　　　　　　Annual maintenance cost
　　　　　　　Fuel, animal feeding or labour cost

Technical: Lifetime of each component

The period of analysis should be at least equal to the economic lifetime of the system. This can generally be taken as the useful lifetime of the longest-lived component (usually the PV array). For consistency, the same analysis period should be used for all the pumping methods being compared.

Discount rate is taken relative to general inflation (ie in real rather than nominal terms), and in most economies is fairly stable at around 10 %. Likewise, inflation rates for various items (eg. parts or fuel) are relative, and so for most items will be zero.

Step 2: Capital costs

The capital costs are the initial costs incurred when the pumping system is first purchased and set up. They tend to depend on component size, and are proportional to the system's rated power to a certain extent (eg. PV modules are around $5 per Wp (f.o.b, 1990 prices).

If you have received a quote for a complete solar pumping system then you can use that as it stands. Otherwise, you can make an estimate from the cost guidance data in section 8.3.1.

Appendices G.1, G.2 and G.3 give cost guidance for windpumps, diesel pumps and hand pumps respectively if no real prices are available locally. The capital cost of wind and diesel pumps will vary with their size, as calculated in the previous step.

As the size of handpumps is not variable, the capital cost will depend on the number of hand pumps required.

To determine the capital cost of the storage tank and distribution systems (where included) a decision must be made on tank volume and pipe or channel size. (Pipe size should already have been estimated from the maximum flow discussion in section 6.5). If local price information is not available, estimated figures can be found in section 8.3.1.

The capital cost figure should include an allowance for transportation and installation, and any other civil works such as a fenced enclosure or engine house... Installation should include the cost of labour, raw materials, and the drilling of a borehole or well if necessary. Remember that if more than one handpump is necessary you will probably need more wells.

100

Step 3: Recurrent costs

Recurrent costs can be considered to consist of three parts: replacements, maintenance and operation. The sections 8.3.2 and 8.8.3 on costs guidance give advice and figures for these for solar power. Appendix G gives similar data for wind diesel and handpumping.

Replacement costs, occurring at intervals corresponding to the estimated lifetime of each component. For instance, if the analysis period is fifteen years, and, say, an inverter is estimated to last five years, then replacements will occur at years 5 and 10.

Each replacement cost must be multiplied by the appropriate factor Fs, as given in table F.1 of appendix F, and as described in the example in section 8.1.2(a).

Maintenance and repair costs, occurring each year. These will vary depending on the system used, and the cost of labour. In general, the more complex the system, the higher the maintenance costs. Solar pumping systems require very little maintenance, and this can usually be carried out by the users. However a major problem will probably require a trained engineer, which could be expensive. On the other hand, a diesel pump may need regular servicing, but as diesel technology is widespread, an engineer will probably be locally available at fairly low lost. The maintenance tasks needed for a solar pump are outlined in section 9.4.

In the absence of real prices, the approximate yearly maintenance for the different systems can be taken as follows:

Solar 1% of installed capital cost (hardware) per year Wind 2% Diesel $200/year Hand 15%/year of installed capital cost.

When calculating the present worth of the sum of the contributions from the maintenance & repair costs, take inflation as zero unless labour costs are expected to change relative to the general inflation rate.

Operating costs: These may be fuel costs or labour charges for operation and attendance. Most solar borehole pumps will not need an attendant, but portable floating pumps in shallow surface water are in danger of running dry if left permanently unattended. The operating costs are therefore dependent upon the price of labour and the pumping situation.

Wind pumps can be assumed to have zero operating costs.

For diesel pumps the main costs are the fuel, calculated on the basis of hours of operation. Figure G.2 in Appendix G.2 shows a graph by which hours (per day) of operation, and therefore fuel costs, can be calculated for a 2.5 kW engine consuming 1.5 l/h. The annual fuel costs are then found by multiplying the hours per year by the cost per hour. It should be noted that fuel costs may vary considerably with location: If this is being performed as an economic analysis rather than a financial

101

one, then the international price should be used (around $0.20 to $0.30 per litre). This is markedly different to local prices, which in remote areas can be as high as $2.00 per litre.

The operating costs for a hand pump, attributable to wages, are obtained by multiplying the number of humans by the daily labour charge rate. From this the annual labour cost can be found. If it is anticipated that labour costs are going to change in real terms, then the relevant inflation rate should be used when finding Fa from table F.2 in appendix F.

Because maintenance and operating costs are annually recurring, they must be multiplied by the appropriate discount factor Fa (from table F.2, appendix F) to give the total present net worth. (As described in the example in section 8.1.2(b)).

If being used in a 'cash-flow' style financial analysis (see discussion in section 8.1.1), recurring costs should be used at their 'year zero' value (ie without discount factors).

Step 4: Life-Cycle Costs

We have now calculated everything necessary to find the total life-cycle cost for each system over the whole analysis period. This is the sum of the capital costs from Step 2, and the total recurrent costs (expressed in terms of total present worth) from Step 3. The relative sizes of the Life-Cycle Costs (LCC's) for each system gives an indication of which is the most viable in economic terms (ie the most economic system has smallest LCC).

The process could end at this point, but the figures we have do not tell us much in real terms. For convenience, there are two other ways of expressing the life-cycle costs that are perhaps more meaningful and easily understood.

Annualised Life-Cycle Cost: Essentially, we are expressing the total LCC in terms of a cost per year. However, we cannot simply divide the total LCC by the number of years in the analysis, as this takes no account of the changes in value of money due to discount rates. The total LCC must be divided by the factor Fa from table F.2 in appendix F, found using the chosen discount rate, inflation rate of zero, and a number years equal to the analysis period. This is really the reverse process of discounting, and the result is expressed in $/year for each system.

Unit water costs: Finally we can convert the annualised life-cycle cost (or ALCC) to a unit water cost. This tells us the true cost of the water provided in terms of $/m3. This is found by first calculating the total volume of water usefully pumped in a year: For each month, multiply the 'actual daily volume pumped' columns (as per examples in tables 6.4 and 7.3) by 30 to find the approximate volume pumped per month. Then sum these monthly totals to find the annual total. Dividing the ALCC by the annual water demand will give the effective unit cost of the water in $/m³. If the water volume pumped is very much surplus to requirements for some months, and will not be used (a most unlikely scenario), then use the real estmated daily water use rather than the actual volume pumped to calculate the annual pumped volume.

The life-cycle costing process may appear rather long and complex, but if a little care is exercised it provides a relatively straight-forward way to make a valid economic comparison between different options. In consideration with other non-economic factors this forms a vital part of the decision making process.

An example of a life-cycle cost comparison like that described above is given in appendix H. The example is for a village water supply situation with the water table 5m below ground level, and a village population of 500.

8.3 Guidance on Costs

This section is intended to give some rough guidance on the costs of various components of a typical solar pumping system. Naturally these will vary somewhat between suppliers, but for a general life-cycle cost analysis the prices given below will provide the reader with an idea of what to expect. Some of the figures give estimates of non-material items such as installation, maintenance and operating costs. In these cases, costs will depend very much on local conditions, and so local data should be used in preference where available. However, in the absence of any necessary piece of real data, the information below should allow the reader to perform a complete estimated life-cycle cost analysis for a solar pumping system. Summarised cost guidance for wind, diesel and hand pumping are included in appendices G.1, G.2 and G.3 respectively.

8.3.1. System Capital Cost

As manufacturers tend to quote for complete systems it is more representative to use the whole system price for the capital cost. This can be expressed as a function of the hydraulic duty, or also to some extent as a function of the array rating in Wp. It is perhaps easier to estimate a whole system price rather than trying to sum component prices, as these can vary quite markedly. For instance, a more efficient motor may cost more, but requires fewer PV modules, and so total system prices tend to balance out.

Using 1990 prices it has been found that total pumping system capital costs vary between 10 and 50 $ per m^4 or 8 and 30 $ per Wp array rating. The graphs in figure 4.3 and 4.4 are compiled from current (1990) manufacturers literature, and illustrate the total system capital cost as a function of both rated Wp and hydraulic duty for a range of different pump makes and types.

In this section two ways are presented to estimate the capital cost for a prospective system. It can either be found as a function of the array rating in Wp, or of the hydraulic duty in m4/day. This is summarised in tables 4.2(a) and 4.2(b) in the chapter on available equipment, and is derived from the most recent manufacturers price lists for a sample of more than 120 different systems (see section 4.3.2 and appendix B). Although summarised here for convenience, you may prefer to use the data as shown in full in graphical form for each configuration in appendix E.

As smaller systems tend to be more expensive on a $/Wp or $/m_4/day basis, prices for two power ranges have been given for each pump configuration. In general a better correlation is found between $/m^4/day and hydraulic duty than between $/Wp and array Wp except for suction pumps. It is therefore preferable to use table 4.2(b) where possible for submersibles, floating pumps and jack pumps, and table 4.2(a) for surface suction pumps. However, remember that the system sizes are usually defined for an insolation of 6 kWh/m$_2$ in the manufacturers specifications, and so if the design month insolation is significantly below this it will be better to use table 4.2(a) for all the systems.

For each configuration and power range, a range of prices is given. This represents the approximate spread of the real price data, and so the centre of this range should give you a realistic estimate of the capital cost. When calculating costs in this way, bear in mind that none of the systems retailed at lower than about $1500. Therefore use $1500 as a minimum figure and ignore any calculated system cost that works out below this.

On top of the pumping system capital cost you will probably also want to consider the costs of storage and pipework, and also the cost of installation. When a system is purchased it should already include the necessary plumbing accessories and support structure, but the user will have to supply any storage or additional pipework associated with the distribution system. If local prices for these are not available then the following figures can be used as a guideline :

Extra cable (if necessary)	5	$/m
Standard pipework	5	$/m
Storage tank cost	60 to 150	$/m³
Header tank * filter	$100 - $200	(irrigation only)

Installation costs will vary with the local cost of labour for a suitable engineer, but will also include a component for consumables. Some types of pumps will cost more to install than others. For instance a floating pump needs practically no installation except for the array, whereas a jackpump may require lifting apparatus to position the down-the-well components. Borehole drilling is expensive and may well turn out to be a significant part of the total system cost. To estimate this it is necessary to know the required depth of the borehole and the cost-per-metre of borehole drilling. If this is not known 60 $/m can be taken as a typical figure (so a 20m borehole would cost $1200).

Borehole drilling (if necessary)   60 to 200 $/m

Pump & Power installation:
Borehole Submersible $250 + 0.6 $/Wp

| Surface or floating | $100 + 0.6 $/Wp |
| Jack pump | $650 + 0.6 $/Wp |

If installation is undertaken by the user, then these will be reduced accordingly.

In the case of irrigation system it is also necessary to account for the cost of the field application system. The two most appropriate systems are low-head drip and hose & basin. The first of these, the low-head drip is the more advanced and more efficient system, costing around $2500 per irrigated hectare. The hose & basin system is a much cheaper solution (although more labour intensive) at, say, $100-$200 per hectare.

8.3.2 Replacement costs and component lifetimes

This section gives the capital costs of the various items that will probably need replacement during the lifetime of the system. The system lifetime is taken as the lifetime of the PV array (estimated at 20 years), as this is the longest-lived component, and so we will not need to consider its replacement value. It must be remembered that all future replacement costs must be discounted to the present net worth as described in section 8.1.2(a). For solar pumping the only items that you will probably need to consider for replacement are the motor/pump unit and the power control electronics (if any).

Component costs derived from the manufacturers data are also detailed in section 4.4 in the chapter on available equipment.

The most recent manufacturers price data for motor/pump replacements has been analysed for the different pump types:

Submersibles: In general 1 - 3 $/Wp of the array rating. Actual prices between $1000 and $3000.

Floating: Around $ 1000

Jack pumps: In general 10 - 15 $/Wp, but not lower than about $3000.

Surface suc.: Between $1000 and $1500.

Lifetimes for most motor/pumps are estimated at around 7 to 10 years.

Other items:

Component	Price	Estimated lifetime
Array	6 - 10 $/Wp	System Life (20 years)
Support structure (add 4 $/Wp for tracking array support)	1 - 2 $/Wp	System life (20 years)
Pipework/storage		System life (20 years)
Inverter	$1000 - $2000	10 years
Simple power control	$450	10 years
Batteries (if any)	See note below	5 years

It is not usual to use batteries with solar pumping systems, and so they have not been included in our discussions. However, a few manufacturers do include them. If you are performing the life-cycle analysis for such a system that you have been offered, use the supplier's replacement battery price. If normal lead-acid automotive batteries are specified, use the local battery price, or use 150 $/kWh of battery capacity.

8.3.3. Maintenance and Operation Costs

Annual maintenance costs are very low for solar pumps, and can be estimated at about 1 % (per year) of the installed system capital cost.

Operating costs for solar pumps are usually zero as they need no attendant either for startup or during their daily working period. As some intervention may be necessary for irrigation pumping, a figure of 1 to 2 $/man-day can be used. If a 'water committee' exists in the case of a jointly owned village supply system it may be necessary to pay someone a small retainer for routine maintenance, collection of payments etc. Unless actual figures are known, this is probably best ignored for the sake of simplicity.

Both of the above are dependent on the local cost of labour, and both are annually recurring and so should be discounted to their present worth using the appropriate discount factors as described in section 8.1.2(b).

9. IMPLEMENTATION AND OPERATION

9.1. System selection

The selection of the appropriate type of pumping system for your situation should be the culmination of all the assessments, evaluations and decisions outlined in the previous chapters. By this point, you should have:

- Familiarised yourself with solar pumping technology and terminology from sections 2. and 3.

- Performed a site assessment and estimated the size of the system components from sections 6. or 7.

- Performed a life-cycle cost analysis from section 8., to compare solar pumping with wind, diesel and hand powered systems.

If you are missing some piece of information that prevents you from completing the above steps, then a sensible estimate will usually suffice, and probably not make too significant a difference to the final result. It should be stressed that the methods outlined in this booklet will only give approximate results, and are designed to point the potential user/buyer in the right direction rather than define an exact solution. This applies particularly to the data in the appendix on wind, diesel and hand pumping (appendix G.): These rough estimates are included only for the purposes of performing the life-cycle cost analysis, and should on no account be used to select a particular wind, diesel or pumping solution without further reading (see Bibliography).

If you have satisfied yourself that solar pumping is both technically feasible and economically viable for your situation, you must now look very carefully at the other non-technical, non-economic factors that may influence your decision. The importance of these factors will vary with locality, but some main questions to consider are:

- **Diesel price and supply security** - If solar and diesel powered systems are both comparable in terms of cost, then it may be worth considering what will happen to diesel prices in the future. Reliable diesel availability and fuel quality may also be a problem in rural areas. Combined with taxation, this can lead to very much higher local than international prices. This can produce a situation in which diesel may be more cost-effective in an economic analysis, but unviable compared to solar or wind in a financial analysis.

- **User acceptance and technical ability** - Communities in many parts of the developing world will probably never have seen PV equipment before, and this unfamiliarity can mean that the new technology is not well accepted. Ultimately, this can mean that when the first problem occurs, the equipment may be left idle while villagers go back to their old methods of water raising. However, this is very seldom a problem, and acceptance has generally been found to be extremely good. User training and education has a great role to play here. A project in Mali set up a 'water committee' in each village, whose responsibility it was to periodically collect payment from villagers and perform simple maintenance and cleaning on the pumping system. User involvement in the planning and running of pumping programmes has been found to be an essential ingredient to their long-term success.

- **Spares and maintenance availability** - All pumping systems require spares or maintenance at some point. For an unfamiliar technology like PV, it is important that there is a trained engineer within a reasonable distance, and that spares can be obtained at least within the country of use. Lower-tech systems like wind or diesel can probably be dealt with by a local engineer, and any unavailable parts fabricated in a local workshop.

- **Security** - Some small solar arrays (particularly those used with floating sets) are designed to be portable, so that farmers may move them from field to field as required. In some regions this may lead to theft if the equipment is left unattended. This applies to some extent to permanent installations as well. The problem is seldom encountered at present, but as PV becomes more popular, the wider applications of PV modules will be realised, possibly leading to an increase in theft of modules.

If possible seek the advice and experiences of others who have used (or considered using) solar pumping in your area. If you think any other factors may be important, it may be possible to discuss these with your supplier before a final decision is made.

9.2. Procurement

Having decided that solar pumping is a viable and cost effective solution to your water supply problem, we may move on to the system procurement.

There are two options in procurement:

(1) Purchase a total system designed by a qualified supplier and have it installed and tested for you.

(2) Purchase a total system designed by a qualified supplier and install and test it yourself.

108

The first of these choices, known as the 'turnkey' system, is recommended for agencies whose field workers have little experience of PV pumping. Although capital costs on a turnkey system may be higher, this can be more than offset by the guarantee of a properly installed and operational system.

The procurement process can be divided into five steps:

(1) Prepare tender documents

A suggested format for tender documents is given in the 'Solar Pumping Handbook' (see bibliography), together with a technical specification format sheet. This should give the supplier all the data needed to offer a suitable system. As much additional information as possible should be listed, including, if available, average and extreme values of ambient air temperature, relative humidity, windspeed, water temperature, water quality and any known details of soil type and well draw-down characteristics. The possibility of sandstorms, hurricane force winds, and other extreme environmental phenomena should also be mentioned. Finally, remember only to specify performance and do not include design choices or equipment selections. Doing so will compromise your position should the unit fail to produce the required water output.

(2) Issue a call for tenders

Letters may be sent to suppliers with a brief description of the required system. Interested suppliers will then reply with a request for tender documents. A list of some of the major PV pumping system suppliers is given in appendix C.

(3) Preliminary evaluation

Each tender received should be checked to ensure that:

(a) The system being offered is complete and includes spare parts and installation and operating instructions.

(b) The system being offered can be delivered within the maximum period specified.

(c) That an appropriate warranty can be provided.

(4) Detailed assessment

A detailed assessment of each tender should be made under the following four headings, with approximately equal importance attached to each heading:

(a) Compliance with specification

The output of the system should be assessed, taking into account any deviations from the specification proposed by the tenderer.

(b) System design

The suitability of the equipment for the intended use should be assessed taking into account operation and maintenance requirements, general complexity, safety features, etc.

The equipment life should be assessed with regard to bearings, brushes and other parts liable to wear and tear.

The content of the information supplied to support the tender should be assessed, in particular the provision of general assembly drawings and performance information.

(c) Capital cost

Capital costs should be compared, allowing for any deviations from the specification proposed by the tenderer. This can be done by comparing the capital cost per m^4 of water delivered by the system.

(d) Overall credibility of tender

The experience and resources of the tenderer relevant to solar pumping technology should be assessed, together with the tenderers ability to provide a repair and spare parts service should problems be experienced with the solar pump. A reasonable warranty, at least regarding spare parts, should be provided.

(5) Place the order

Select the most appropriate solar pump to meet your requirements and budget. Ensure that you are satisfied that solar pumping is the most appropriate use of the available funds and that satisfactory replies have been received from all the manufacturers on all your queries relating to the equipment. Also ensure that all necessary tools and components needed to complete the installation are included in the order, and that spare parts sufficient for five years operation have been included.

9.3. Installation and operation

When installing the equipment make sure that the following points are taken note of:

PV arrays

The array should be sited in a positions such that it is unshaded by buildings or trees between the hours of 0730 to 1630. Bear in mind that the sun changes position with the season and that trees grow! The array should be fenced in to protect it from possible damage by animals, children or vehicles, and sited to protect it from flood damage. Try and reduce cable lengths to a minimum to reduce power losses, but ensure that there is safe access for cleaning. Power control equipment and junction boxes should be shaded and protected from rainwater penetration.

The manufacturers instructions should be followed at all times, and care taken not to lose small parts such as nuts, bolts and washers. All connections should be cleaned before assembly and the array kept covered throughout installation. All joints should be checked for tightness after erection. Ensure that an adequate earth connection is installed where recommended.

It is vital that the array is oriented in accordance with the manufacturers instructions, or the system will not perform as expected.

Submersible motor/pump units

All cable connections and seals should be checked for water tightness if possible, and a supporting wire should be attached to the motor/pumpset if plastic riser pipes are used. Check the direction of rotation in accordance with the manufacturers instructions; Centrifugal pumps will still work in reverse, but less effectively.

Floating motor/pump sets

Seals, cable glands and the flotation device should be checked as appropriate. Do not lift or support the unit by its cable, but attach a rope to a suitable point to lower it onto the water. Allow sufficient cable and hose for changes in water level, and provide hose support if necessary in wells or fast flowing rivers.

Jack and piston pumps

When mounted on an open well, cover the well while working on the pump unit. Ensure that rod and riser-pipe joints are tight. A coarse strainer is fitted on the inlet pipe, but filters should also be fitted on the outlet side where they can be easily cleaned. Adjust balance weights and pulleys in accordance with the manufacturers instructions. The pump should be fenced off or fitted with guards to keep people or animals away from moving parts.

Surface mounted suction pumps

The suction lift should be as small as is practical, with a steady pipe rise to prevent air locks. The priming tank should be filled before use and the inlet freed from sediment or debris. Check that the direction of rotation of the motor is correct.

9.4. Maintenance

Solar pumps need very little maintenance, but neglect of the following tasks will eventually lead to reduced performance and possible system damage.

Array surfaces should be kept clean by wiping with a wet cloth about once a week. Bolts and electrical connections should be checked for tightness periodically, and lubricate moving parts as required.

At regular intervals checks should be made for:

- Water in floating units
- Blocked strainers and filters
- Damage to cables ropes and support wires
- Leaks in suction and delivery hoses
- Non-return valve operation if applicable
- Motor-bushes as recommended
- Operation of safety cut-out if applicable

In addition it cannot be over-stressed that the routine maintenance specified by the supplier/manufacturer must be carried out.

9.5. Monitoring and evaluation

As PV pumping is fairly likely to be a new technology to the region, it is important to obtain performance data to assess the long term technical, economic and social success of the project. The experience gained will be valuable in any future water supply projects undertaken by your organisation. A monitoring programme should include aspects on:

(i) Methods of application: to gain experience of the way farmers and villagers use solar pumps.

(ii) Economics: to monitor the long term real costs and benefits associated with solar pumping.

112

(iii)   User reaction: to find ways, if necessary, in which the systems can be made more acceptable to users.

(iv)   Performance and reliability: to create a data base on system performance, and particularly to record failure modes and obtain estimates of component lifetimes. All users should keep a log in which maintenance and details of any failures are recorded.

Naturally, performance evaluation cannot be carried out routinely for every pump, but a handful of representative systems should be monitored in detail. The simplest method of evaluation is to take daily readings of:

- Global solar irradiation in the plane of the array
- Volume of water pumped
- Static head

From figure 6.2. in section 6.5., the daily hydraulic energy and hence daily sub-system efficiency can be obtained for different values of solar insolation.

Your supplier will be able to advise you about monitoring instrumentation, such as solarimeters, integrating flow meters and water level measuring devices.

Figure 9.1: Village water distribution point

BIBLIOGRAPHY

World Bank/UNDP Bilateral Aid Energy Sector Management Assistance Program. Activity Completion Report No. 108/89, 1989. "Assessment of Photovoltaic Programs, Applications and Markets (Pakistan)". Washington, D.C.

Doorendos, J., and W. Pruitt. 1977. "Crop Water Requirements". FAO, Rome.

Okun, D.A., and W. R. Ernst. 1987. "Community Piped Water Supply Systems in Developing Countries". World Bank Technical Paper No. 60. Washington, D.C.

Terrado, Ernesto N., Matthew Mendis and Kevin Fitzgerald. 1988. "Impact of Lower Oil Prices on Renewable Energy Technologies". Working Paper No. 5. Industry and Energy Department, World Bank. Washington, D.C.

de Maroc, Royaume. "Journees sur les applications du pompage solaire et eolien au Maroc". Project 608-0159. Ministere de l'Energie et des Mines/IDEA for USAID.

Joint report Meridian Corporation/IT Power Ltd., prepared for US Committee on Renewable Energy, Commerce and Trade. 1990. "Learning from Success: Photovoltaic Powered Water Pumping in Mali". Washington, D.C.

McNelis, B. and A. Derrick. 1988. "Photovoltaic Water Pumping: Experience to 1988". Proceedings Euroforum New Energies Congress, Saarbrucken, FRG.

Derrick A., Catherine Francis and Varis Bokalders. 1989. "Solar Photovoltaic Products: A guide for Development Workers". IT Publications.

McNelis, B., A. Derrick and M. Starr. 1988. "Solar Powered Electricity". IT Publications.

Kenna, J. and W. Gillett. 1985. "Solar Water Pumping: A Handbook". IT Publications.

Adkinson and Associates. 1989. Technology Review Report on Photovoltaic Powered Water Pumping. Prepared for Energy, Mines and Resources. Ottawa, Canada.

McNelis, B. 1989. "The Direct Conversion of Solar Energy to Electricity". Status Report. IT Power, prepared for UN/DIESA Committee on the Development and Utilisation of New and Renewable Sources of Energy.

Thomas, M. G. 1987. "Water Pumping: The Solar Alternative". (SAND87-0804). Sandia National Laboratories, Albuquerque.

Fraenkel, P. 1986. "Water Pumping Devices". IT Publications.

Lancashire, S., J. Kenna and P. Fraenkel. 1987. "Windpumping Handbook". IT Publications.

GLOSSARY OF TERMS

Alternating current (A.C.) - Electric current in which the direction of flow is reversed at frequent intervals. Converse of direct current (D.C.).

Amorphous - The condition of a solid in which the atoms are not arranged in an orderly pattern; not crystalline.

Annualised cycle cost (ALCC) - The total life time cost of a system expressed as a sum of annual payments.

Balance of system (BOS) - Parts of a photovoltaic system other than the array.

Borehole - A hole drilled down to the reach water under the ground. Standard borehole diameters are 4 or 6 inches.

Capital cost - Initial financial outlay on a project.

Centrifugal pump - Pump that produces a pressure difference by throwing water outwards from a central impeller.

Clearness index - The ratio of global solar irradiation to extraterrestrial solar irradiation.

Concentrator - A photovoltaic array which includes an optical component such as a lens or focusing mirror to direct incident sunlight onto a solar cell of small area.

Conversion efficiency (cell) - The ratio of the electric energy produced by a solar cell (under full sun conditions) to the energy from sunlight incident upon the cell.

Deep discharge - Discharging a battery to 20 per cent or less of its full charge.

Design month - For the purpose of sizing a solar photovoltaic system, it is necessary to choose a 'worst month' for which the system must meet the load requirements. This month is termed the design month.

Diffuse radiation - Solar radiation scattered by the atmosphere.

Direct radiation - Solar radiation transmitted directly through the atmosphere.

Direct current (D.C.) - Electric current in which electrons are flowing in one direction. Converse of alternating current (A.C.).

Discount factor - Factor by which a future cost or benefit is multiplied to calculate its equivalent present worth.

Discount rate - Rate at which the value of money changes, relative to general inflation. Usually taken as 10% per year.

Drawdown - The distance below the natural water table that the water-level in a well falls to when steady-state pumping is in progress.

Dynamic head - The head loss in pipes caused by the flow of water through the pipes.

Extra-terrestrial irradiation - The solar energy received outside the earth's atmosphere.

Fill factor - The ratio of maximum output of a PV cell under reference conditions to the product of open circuit voltage and short circuit current/under the same conditions.

Flat plate (module or array) - An arrangement of solar cells in which the cells are exposed directly to normal incident sunlight. Opposite of concentrator.

Global irradiance - The sum of diffuse and direct solar irradiance incident on a horizontal surface.

Head - The height of water column that would produce the pressure that the pump experiences.

Hydraulic duty - The work that must be done to lift water, expressed as hydraulic energy, or as a volume-head product.

Hydraulic energy - The energy necessary to lift water.

Impedance matching - The process of matching the output of one device to the input of another device such that there is a maximum transfer of power between the two.

Insolation - Sunlight, direct or diffuse (not to be confused with insulation).

Inverter - Device that converts D.C. electrical power to A.C.

I-V curve - A graphical presentation of the current versus the voltage from a photovoltaic cell as the load is increased from the short circuit (no load) condition to the open circuit (maximum voltage) condition. The shape of the curve characterises cell or module performance.

Jack pump (Nodding donkey) - Reciprocating pump in which the motor is on the surface, and the pump down the well. Used for high head applications.

Kilowatt (kW) - 1,000 Watts.

Kilowatt hour (kWh) - 1,000 Watt hours.

Life Cycle Costs (LCC) - The lifetime costs associated with a pumping system expressed in terms of today's money.

Load - Any device or appliance that is using power.

Maximum Power Point Tracker (MPPT) - Impedance matching electronics used to hold the output of the PV array at its maximum value.

Nodding Donkey - See Jack pump.

Parallel connection - A method of interconnecting two or more electricity-producing devices, or power-using devices, such that the voltage produced, or required, is not increased, but the current is additive. Converse of series connection.

Peak Watt or Watt peak (Wp) - The approximate amount of power a photovoltaic device will produce at noon on a clear day (insolation at 1000 Watts per square metre) when the cell is faced directly toward the sun.

Photovoltaic (PV) - Pertaining to the direct conversion of light into electricity.

Photovoltaic array - An interconnected system of photovoltaic modules that functions as a single electricity-producing unit. The modules are assembled as a discrete structure, with common support or mounting.

Photovoltaic cell - A device that converts light directly into electricity. A solar photovoltaic cell, or solar cell, is designed for use in sunlight. All photovoltaic cells produce direct current (D.C.).

Photovoltaic module - A number of photovoltaic cells electrically interconnected and mounted together, usually in a common sealed unit or panel of convenient size for shipping, handling and assembling into arrays.

Photovoltaic system - A complete set of components for converting sunlight into electricity by the photovoltaic process, including array and balance-of-system components.

Poly-Crystalline silicon; polysilicon - Silicon which has solidified at such a rate that many small crystals have formed. The atoms within a single crystal are symmetrically arrayed, whereas in polysilicon crystals they are jumbled together.

Positive displacement pump - One that moves water by means of a cavity or cylinder of variable size. Also known as a volumetric pump.

Power conditioner - The electrical equipment used to convert power from a photovoltaic array into a form suitable for subsequent use, as in supplying a household. Loosely, a collective term for inverter, transformer, voltage regulator, meters, switches and controls.

Present worth - The value of a future cost or benefit expressed in present day money.

Prime mover - The power source for a system.

Progressive cavity pump - Positive displacement pump that has a helical cavity between the rotor and stator. As rotor rotate the cavity 'progresses' upwards carrying the water with it.

PV - Abbreviation for photovoltaic.

Reciprocating pump - Pump driven by a push/pull motion rather than by rotation.

Rotor - Rotating central section of motor or pump assembly.

Series connection - A method of interconnecting devices that generate or use electricity so that the voltage, but not the current, is additive one to the other. Converse of parallel connection.

Short circuit current - Of a PV cell, module or array is the current that flows when the output terminals of the device are joined together.

Solar irradiance - The power received per unit area from the sun.

Solar irradiation or insolation - The energy received per unit area from the sun in a specified time period. In this handbook, the time period is generally taken to be a day and the solar irradiation is expressed in MJ per m^2 per day or kWh per m^2 per day (1 kWh = 3.6 MJ).

Stand alone - An isolated photovoltaic system not connected to a grid; may or may not have storage, but most stand-alone applications require a battery or other form of storage.

Static head - The height over which water must be pumped, i.e., from the well level to a storage tank.

Stator - Outer, stationary component of pump assembly (see 'progressive cavity pump').

Subsystem - Usually refers to the just the motor/pump set. Does not include the PV array.

Tilt factor - The ratio of solar irradiation incident on a tilted PV array to global irradiation.

Total head - Sum of the static and dynamic heads and drawdown.

Village water supply - Water supply for drinking and other domestic purposes in rural communities. This may also include some livestock watering or micro-irrigation (i.e., household vegetable garden). Referred to in some publications as 'rural water supply'.

Volume-head product - The daily water volume multiplied by the total head - used as a measure of the hydraulic duty, as it is proportional to the hydraulic energy. Units are m^4.

Watt, wattage - A measure of electric power, or amount of work done in a unit of time. One amp of current flowing at a potential of one volt produces one Watt of power.

Watt hour (Wh, Whr) - A quantity of electrical energy (electricity). One Watt hour is consumed when one Watt of power is used for a period of one hour.

APPENDIX A: GLOBAL DAILY INSOLATION MAPS

The following maps are included for use when information regarding local insolation is not available.[1] The author would like to acknowledge the cooperation of Sandia National Laboratories, USA in the reprinting of these maps. The seasons mentioned in the titles for each chart are defined for the northern hemisphere. In the southern hemisphere the seasons will be reversed.

The tilt angle is defined as the angle at which the array is raised from the horizontal in order to capture the suns rays most effectively. A tilt angle of latitude-angle plus 15° has been used in these maps, to give improved sunlight collection at low sun angles. The tilt angle is measured with the array pointing southwards in the northern hemisphere and northwards in the southern hemisphere.

---

[1] The author would like to acknowledge the cooperation of Sandia National Laboratories, USA in the repringint of these maps.

Figure A-1. Insolation Availability (Latitude + 15°
Tilt, Winter)

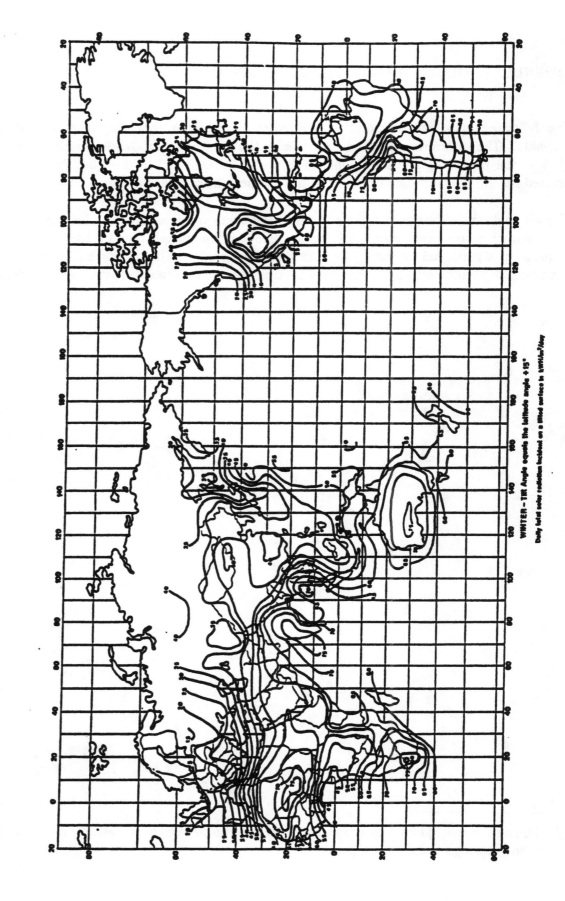

WINTER – Tilt Angle equals the latitude angle +15°

Daily total solar radiation incident on a tilted surface in kWH/m²/day

Figure A-2. Insolation Availability (Latitude + 15° Tilt, Spring)

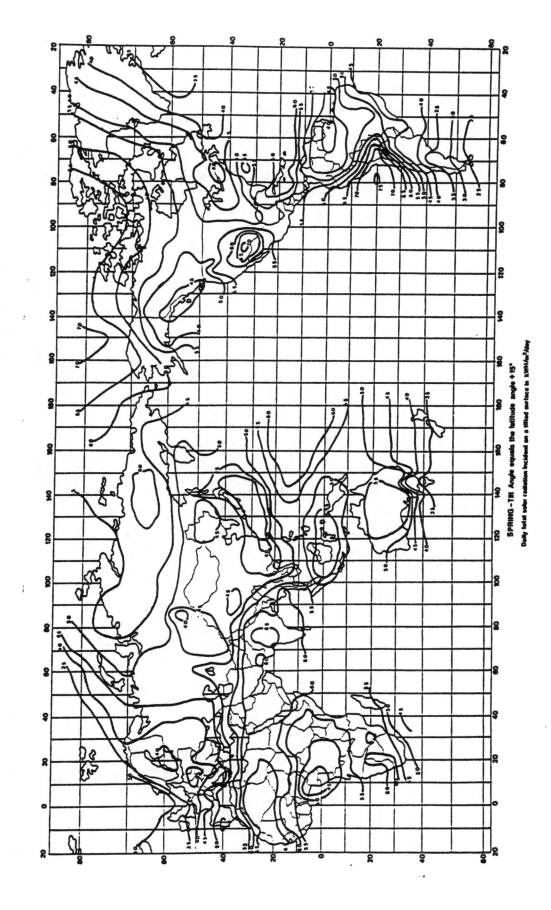

SPRING-Tilt Angle equals the latitude angle + 15°

Daily total solar radiation incident on a tilted surface in kWh/m²/day

.Figure A-3. Insolation Availability (Latitude + 15°
Tilt, Summer)

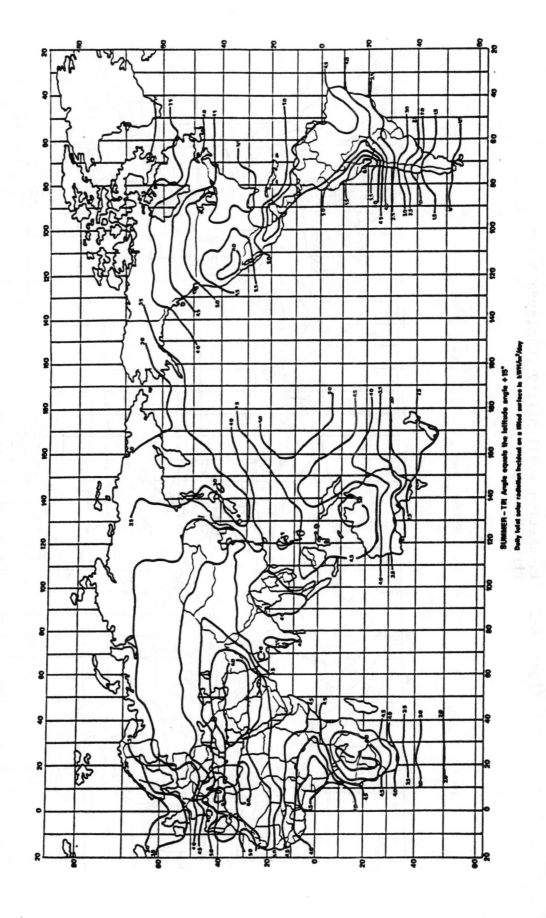

SUMMER – Tilt Angle equals the latitude angle +15°

Daily total solar radiation incident on a tilted surface in kWh/m²/day

Figure A-4. Insolation Availability (Latitude +15°
Tilt, Autumn)

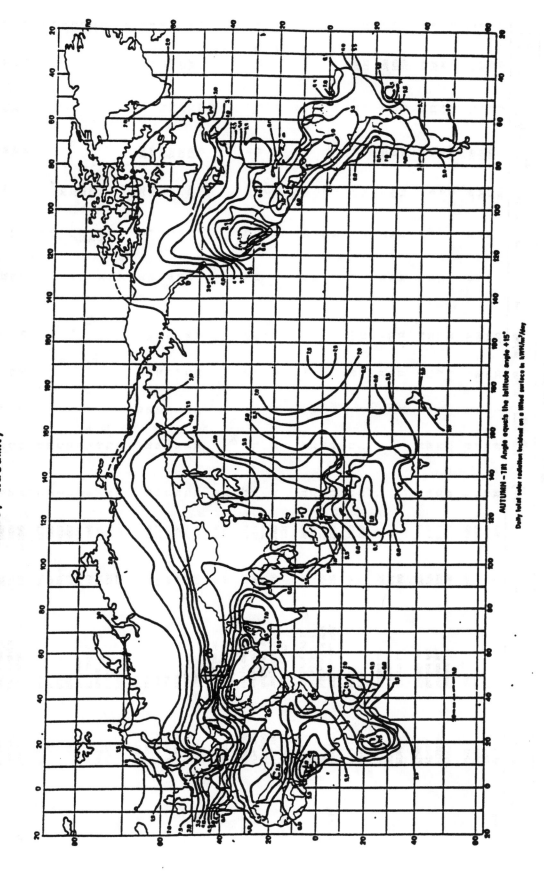

AUTUMN – Tilt Angle equals the latitude angle +15°

Daily total solar radiation incident on a tilted surface in kWH/m²/day

APPENDIX B: SUMMARY OF MANUFACTURERS DATA

Table B.1: Manufacturer's product data

Manufacturer	Model	Type	Panel WP	FOB Cost	Cost $/Wp	Cost $/m⁴	Low Head	m³/day	m⁴/day	Mid-range Head	m³/day	m⁴/day	High Head	m³/day	m⁴/day	Solar Irrad.
BP Solar	SP14	Sub	560	7870	14	26	5	30	150	20	15	300	65	1	65	6
	SP21	Sub	840	10170	12	18	5	40	200	30	19	570	100	2	200	6
	SP28	Sub	1120	12460	11	18	5	50	250	30	23	690	100	5	500	6
	SP35	Sub	1400	14760	11	19	5	55	275	30	26	780	100	6.5	650	6
	PD40-2	Surface suc	80	2460	31	82	5	4	20	10	3	30	20	2	40	6
	PD40-3	Surface suc	120	2790	23	46	5	5	25	20	3	60	40	1.6	64	6
	PD40-4	Surface suc	160	3115	19	39	5	5.2	26	20	4	80	40	2.5	100	6
	PD40-5	Surface suc	200	3440	17	38	5	5.5	27.5	20	4.5	90	40	3.2	128	6
	PD60-9	Surface suc	360				5	15	75	30	13	390	60	6	360	6
	PD60-15	Surface suc	600				5	27	135	30	23	690	60	10	600	6
	PD60-18	Surface suc	720				5	31	155	30	27	810	60	13	780	6
	FP10-8	Float	320	4920	15	27	5	40	200	6.5	28	182	8	19	152	6
	FP10-12	Float	480	6230	13	17	5	68	340	8	46	368	10	21	210	6
Chloride	PFI/1	KSB Aquasol 100L	318	5390	17	45				2	60	120				
	PFI/3	KSB Aquasol 50M	636	8028	13	19				7	60	420				
	PF3/2	Grundfos SP4-8	742	9377	13	23				20	20	400				
	PS2/4	Grundfos SP2-18	1484	15965	11	18				40	22	880				
Chronar	MPL 120	surf pr cav/dc	160	2271	16	23	3	20	60	9	12	108	15	3	45	6
	MPL 160	surf pr cav/dc		2492												
	MPL 200	surf pr cav/dc	240	2828	13	17	5	20	100	15	12	180	23	3	69	6
	MPL 240	surf pr cav/dc	320	3074	12	17	8	20	160	18	12	216	30	3	90	6
	MPL 320	surf pr cav/dc		3747												
	MP 320	surf pr cav/dc	320	3617	11		10									
	MP 480	surf pr cav/dc	480	4796	10		15									
	JP 502	Jack	160	5181	32	32	10	7.5	75	40	4	160	60	2.5	150	6
	JP 753	Jack	240	6460	27	26	15	12	180	50	5	250	120	2.5	300	6
	JP 754	Jack	320	7600	24	22	15	16	240	50	7	350	120	3.5	420	6
	JP 756	Jack	480	10084	21	21	20	23	460	70	7	490	240	2.5	600	6
	JP 1008	Jack	640	12281	19	19	20	25	500	80	8	640	300	2.5	750	6
	JP 1009	Jack	720	13970	19	19	20	25	500	80	9	720	300	3	900	6
	JP 1012	Jack	960				20	25	500	90	12	1080	300	4	1200	6
Dinh	HP500	Piston	100	1788	18	36				26	1.9	50				5
	HP900	Piston	150	2340	16	27				26	3.4	88				5
	3000	Piston	525	4740	9	16				26	11.4	300				5
	8000	SS Centri	525	4740	9	39				4	30.2	120				5
	12000	SS Centri	800	5940	7	24				5.5	45.4	250				5
	S-1000	Sub	975	7692	8	20				100	3.8	380				5
	S-1600	Sub	1625	9132	6	15				100	6	600				5
	PJ1000	Jack	975	6792	7	18				100	3.8	380				5
	PJ1600	Jack	1625	8711	5	15				100	6	600				5
Duba	Minisol	surf suc dc	120	2840	24	47	5	6	30	15	4	60	30	2.3	69	6
	Minisol	surf suc dia/dc	160	3480	22	43	5	7.5	38	15	5.3	80	30	3.5	105	6
	Minisol	surf suc dia/dc	200	4000	20	44	5	8.5	43	15	6	90	30	4.5	135	6
	Supersol	sub mot	640	11438	18	22	3	35	105	25	21	525	40	6	240	6
	Supersol	sub mot	960	16074	17	22	8	35	280	35	21	735	60	5	300	6
	Supersol	sub mot	1280	19034	15	20	13	35	455	45	21	945	70	4	280	6
Fluxinos	Pulsa	Oscillation	360	7407	21	41	10	9	90	30	6	180	50	4	200	6.5

Table B.1. (cont'd): Manufacturer's product data

Manufacturer	Model	Type	Panel WP	FOB Cost	Cost $/Wp	Cost $/$m^4$	Low Head	m^3/day	m^4/day	Mid-range Head	m^3/day	m^4/day	High Head	m^3/day	m^4/day	Solar Irrad
Grundfos	SP 1-28	sub ac	1200	14190	12	24	80	7.5	600	100	6	600	120	4.5	540	6
	SP 2-18	sub ac	1000	13000	13	22	30	20	600	55	11	600	80	5	400	6
	SP 4-8	sub ac	880	10650	12	17	10	47	470	25	25	625	40	10	400	6
	SP 8-4	sub ac	1000	13000	13	19	8	75	600	14	50	700	20	30	600	6
	SP 16-2	sub ac	1000	13000	13	18	5	130	650	8	90	720	11	60	660	6
	SP 27-1	sub ac	1200	14190	12	17	5	190	950	6.5	130	845	8	90	720	6

Heliodinamica

Supply 50 standard systems for heads 5 to 125 m and flows 130 m^3 per day. Power range 350 to 2250 Wp.

Manufacturer	Model	Type	Panel WP	FOB Cost	Cost $/Wp	Cost $/$m^4$	Low Head	m^3/day	m^4/day	Mid-range Head	m^3/day	m^4/day	High Head	m^3/day	m^4/day	Solar Irrad
	SB15	surface	175	4555	26	35				5.5	24	132				
	SB26	surface	420	8211	20	20	10	40	400	13	32	402	15	27	405	
	SB27	surface	490	9247	19	21	12	38	456	15	30	444	18	24	432	
	SB37	sub	735	11624	16	22	14	39	546	26	20	520	38	13	494	
	SB46	sub	840	13366	16	24	14	42	588	34	17	564	54	10	540	
	SB47	sub	980	14982	15	25	14	45	630	36	17	605	58	10	580	
	SB56	sub	1050	15790	15	24	15	45	675	39	17	652	63	10	630	
	SB57	sub	1225	17810	15	21	21	45	945	46	18	832	72	10	720	
	SB66	sub	1260	18214	14	23	19	45	855	46	17	792	73	10	730	
	SB67	sub	1470	20638	14	24	20	45	900	51	17	865	83	10	830	
	SBG-1		17	546												
	SBG-2		40	680												
	SBG-3		80	1084												
	SBG-4		120	1488												
	SBG-5		160	2520												

Helios

Also produce many other systems of intermediate power. KTSP xy-z : xx is Wp of modules, y is no. of modules, y is no. of sets of 8, z is head range A to F.

Manufacturer	Model	Type	Panel WP	FOB Cost	Cost $/Wp	Cost $/$m^4$	Low Head	m^3/day	m^4/day	Mid-range Head	m^3/day	m^4/day	High Head	m^3/day	m^4/day	Solar Irrad
	KTSP 402-B	sub ac	640	17217	27	54	5	80	400	8	40	320	11	10	110	6.5
	KTSP 402-D	sub ac	640	17217	27	43	8	40	320	20	20	400	40	3	120	6.5
	KTSP 453-A	sub ac	1080	23323	22	27	8	68	544	5	175	875	8	54	432	6.5
	KTSP 453-D	sub ac	1080	23323	22	25	80	7	560	20	47	940	40	13.5	540	6.5
	KTSP 453-F	sub ac	1080	23323	22	23				100	10	1000	120	2.7	324	6.5
	KTSP 455-A	sub ac	1800	32895	18	21	8	88	704	20	310	1550	8	189	1512	6.5
	KTSP 455-D	sub ac	1800	32895	18	22	80	13	1040	100	74	1480	40	34	1360	6.5
	KTSP 455-F	sub ac	1800	32895	18	17				100	18.8	1880	120	8.1	972	6.5
	KTPA	surf sub	270	5797	21	95	8	5.4	43	16	3.8	61	30	0.9	27	6
	KTPI	sub dc	270	5411	20	82	4	18	72	6	11	66	7.4	3.6	27	6
Hydrasol	HS-300-1	Sub dc	300	5870	20	36	10	16	160	15	11	165	20	8	160	5
	HS-300-2	Sub dc	300	5870	20	46	3	30	90	8	16	128	13	10	130	5
	HS-300-3	Sub dc	300	5870	20	43	3	31	93	6	23	138	9	15	135	5
	HS-900-1	Sub dc	900	12426	14	24	15	28	420	32	16	512	45	12	540	5
	HS-900-2	Sub dc	900	12426	14	30	7	40	280	17	24	408	27	14	378	5
	HS-900-3	Sub dc	900	12426	14	31	4	51	204	12	33	396	20	15	300	5
	HF-300	Floating dc	300	4328	14	16	23	115	264	2.9	92	267	3.5	76	266	5
	HF-450	Floating dc	450	5848	13	14	3	138	414	4	104	416	5	83	415	5
	HF-800	Floating dc	800	9446	12	11	5	164	820	6	137	822	7.5	109	817	5
	HF-1000	Floating dc	1000	11524	12	12	5	158	790	7	135	945	9	103	927	5
	HF-1400	Floating dc	1400	16204	12	11	7	212	1484	9	166	1494	11	136	1496	5
KSB	Aquasol 50M	floating	530	8570	16	24	2	115	230	6	60	360	10	10	100	5
	Aquasol 100L	floating	530	3480	15	22	1	240	240	3	120	360	4.5	10	45	5
	Coar 100	sub 4"														

Table B.1. (cont'd): Manufacturer's product data

Manufacturer	Model	Type	Panel WP	FOB Cost	Cost $/Wp	Cost $/m⁴	Low Head	m³/day	m⁴/day	Mid-range Head	m³/day	m⁴/day	High Head	m³/day	m⁴/day	Solar Irrad.
Kyocera	SMB-50	Sub dc	50	1760	35	63	2	10	20	4	7	28	6	4	24	6
	SMB-50	Sub dc	50	1760	35	63	5	6	30	7	4	28	11	2	22	6
	SMB-200	Sub dc	200	5105	26	29	1	90	90	2.5	70	175	4	50	200	6
	SMB-200	Sub dc	200	5375	27	34	5	33	165	8	20	160	13	8	104	6
	SMD-750	Sub dc	750	12555	17	15	5	140	700	9	90	810	13	45	585	6
	SMD2-750	Sub dc	750	12555	17	18	10	50	500	23	30	690	39	15	585	6
	SMA-550G	Sub ac	550	9460	17	16	15	30	450	30	20	600	45	10	450	6
	SMA2-550G	Sub ac	550	9460	17	24	2	60	120	10	40	400	20	20	400	6
	SMA-1.5G	Sub ac	1500	23460	16	16	45	25	1125	100	15	1500	140	5	700	6
	SMA4-1.5G	Sub ac	1500	23460	16	16	4	200	800	10	150	1500	15	100	1500	6
	SMA2-3.7G	Sub ac	3700	46920	13	14	70	40	2800	130	25	3250	180	15	2700	6
	SMA4-3.7G	Sub ac	3700	46920	13	10	25	200	5000	35	130	4550	50	70	3500	6
McDonald	8100	Surf suc	130	2830	22	141	3	7	21	5	4	20	7	2.1	15	5
	8100	Surf suc	215	4270	20	102	3	12	36	6	7	42	10	1.3	13	5
	8200	Surf suc	636	10442	16	25	5	65	325	10	42	420	15	31	465	6
	8200	Surf suc	1060	16864	16	21	5	106	530	10	80	800	15	59	885	6
	1500	Surf suc	420	7422	18	53	10	25	250	20	7	140	30	0.7	21	6
	1500	Surf suc	1680	25696	15	33	20	45	900	25	31	775	30	22	660	6
	1800	Sub mot	318	8806	28	49	15	16	240	30	6	180	60	0.7	42	6
	1800	Sub mot	848	16833	20	34	15	40	600	50	10	500	90	2.6	234	6
	1800	Sub mot	1590	28071	18	35	15	59	885	80	10	800	150	2.6	390	6
Mono/Suntron - Produce about 70 different configurations for varying head and power specifications, with racking and stationary arrays.	XMS02BP	Surf mot Stat	120	4500	38	87	5	4.2	21	20	2.6	52	30	1.8	54	7
	XMS04BP	Surf mot Stat	240	4674	19	33	5	15	75	33	4	140	63	1.5	97	7
	XMS08BP	Surf mot Stat	480	7280	15	18	5	35	175	45	9	405	95	1.6	152	7
	XMS12BP	Surf mot Stat	720	9808	14	15	5	52	260	50	13	650	100	3	300	7
	XMS16BP	Surf mot Stat	960	12644	13	14	15	68	340	50	18	900	100	4.8	480	7
	XMS21BP	Surf mot Stat	1260	15939	13	13	15	76	1140	55	23	1265	100	7.6	760	7
	XMS28BP	Surf mot Stat	1680	20536	12	12	25	72	1800	60	29	1740	100	11	1100	7
	XMS02SP	Surf suc	120	4400	37	77	5	11	55	15	3.8	57	25	2.2	55	7
	XMS04SP	Surf suc	240	5900	25	42	5	25	125	35	4	140	60	2.2	132	7
	XMS08SP	Surf suc	480	8812	18	21	5	53	265	50	8.5	425	100	2.1	210	7
	XMS12SP	Surf suc	720	9732	14	13	5	80	400	50	15	750	100	4.5	450	7
	XMS16SP	Surf suc	960	12644	13	14	10	75	750	55	17	935	100	6.6	660	7
	XMS21SP	Surf suc	1260	15785	13	11	15	79	1185	55	25	1375	100	8.1	810	7
	XMS28SP	Surf suc	1680	20460	12	11	35	59	2065	65	29	1885	100	15	1500	7
Siemens: Borehole uses Grundfor pumps	SP 1-28	sub ac	770	9200	12	38	80	3.5	280	100	2.4	240	120	1.2	144	6
			1155	12230	11	22	80	7	560	100	5.5	550	120	4	480	6
			1540	14640	10	18	80	8	512	100	8	800	120	9.6	1152	6
	SP 2-18	sub ac	770	9190	12	28	30	6	420	55	6	330	80	1	80	6
			1155	11620	10	16	30	13	690	55	13	715	80	7	560	6
			1540	14040	9	14	30	18	840	55	18	990	80	11	880	6
	SP 4-8	sub ac	770	8530	11	16	10	27	250	20	27	540	40	7	280	6
			1155	10960	9	13	10	33	550	25	33	825	40	17	680	6
			1540	13370	9	12	10	45	270	25	45	1125	40	63	2520	6
	SP 8-4	sub ac	770	8280	11	17	8	35	496	14	35	490	20	17	340	6
			1155	10710	9	17	8	62	680	14	59	826	20	37	740	6
			1540	13060	8	12	8	85	824	14	79	1106	20	56	1120	6
	SP 16-2	sub ac	770	8220	11	17	5	103	525	8	60	480	11	30	330	6
			1155	10650	9	13	5	140	700	8	105	840	11	75	825	6
			1540	13060	8	12	5	170	850	8	50	1120	11	110	1210	6
	SP 27-1	sub ac	770	8160	11	23	5	105	525	6.5	130	325	8	15	120	6
			1155	10590	9	13	5	180	900	6.5	50	845	8	80	640	6
			1540	13000	8	11	5	235	1175	6.5	180	1170	8	140	1120	6

Table B.1 (cont'd): Manufacturer's product data

Manufacturer	Model	Type	Panel WP	FOB Cost	Cost $/Wp	Cost $/$m^4$	Low Head	m^3/day	m^4/day	Mid-range Head	m^3/day	m^4/day	High Head	m^3/day	m^4/day	Solar Irrad.
Siemens (cont'd)																
(Surface units use KSB Aquasol)																
	50M/1055	Float	550	5410	10	15	2	115	230	6	60	360	10	10	100	5
	50M/1555	Float	825	6850	8	15	1	240	240	3	120	360	4.5	10	45	5
	100L/1055	Float	550	5530	10											5
	100L/1555	Float	825	7000	8											5
SolarJack	SJG/2-53	Jack	106				9	9	81	30	2.5	75	120	0.6	72	5
	SJG/4-53	Jack	212				9	17	153	45	3.5	157	180	0.8	144	5
	SJG/5-53	Jack	265				11	17	187	50	3.5	175	180	1.1	198	5
	SJG/6-53	Jack	318				14	17	238	55	4	220	180	1.3	234	5
	SJG/8-53	Jack	424				18	17	306	60	5	300	180	1.7	306	5
	SDS/1-47	Sub/dc pos.dis	47				1.5	1.1	2	30	0.6	18	70	0.4	28	5
	SDS/2-47	Sub/dc pos.dis	94				1.5	2.5	4	30	1.6	48	70	0.7	49	5
Total Energie	Submersible	Sub ac	640				10	50	500	25	23	575	50	12	600	6
	Submersible	Sub ac	960				10	65	650	25	30	750	55	15	825	6
	Submersible	Sub ac	1280				10	95	950	30	35	1050	65	15	975	6
	Submersible	Sub ac	1600				15	110	1650	35	35	1225	75	17	1275	6
	TPF 240	Floating	240				5	32	160	7.5	24	180	10	14	140	6
	TPF 640	Floating	640				5	80	400	10	40	400	18	26	468	6
	TPF 1280	Floating	1280				5	170	850	10	85	850	15	60	900	6
	TPF 5120	Floating	5120				5	700	3500	10	320	3200	15	230	3450	6

APPENDIX C : LIST OF MAJOR MANUFACTURERS AND DISTRIBUTORS

BHEL
PV Division
Vikasnagar
Mysore Road
INDIA

BP Solar International
36 Bridge Street
Leatherhead
Surrey
KT22 8BZ, UK

BP Thai Solar Corporation Ltd
101/47/9 Nava Nakorn's Ind.
Estate
Phaholyothin Road, Klong 1
Klong Luang
THAILAND

CEL
4 Industrial Area
Sahibabad 201 010
UP
INDIA

Chloride Solar Ltd
Lansbury Estate
Lower Guildford Road
Knaphill
UK

Dinh Compnay
Box 999
Alachua
Florida 32615
USA

Duba s.a.
Nieuwstraat 31
B-9200
Wetteren
BELGIUM

Fluxinos
58100 Grosseto
Via Genova 8
ITALY

Grundfos
DK-8850
Bjerringbro
DENMARK

Heliodinamica
Caixa Postal 8085
9051 Sao Paulo - SP
BRAZIL

Helios Technology
Via PO 8
Galliera 1-35015
Veneta (PD)
ITALY

Hydrasol
Industriestrasse 100
6919 Bammental
GERMANY

IBC
P O Box 1107
D-8623 Staffelstein
GERMANY

Intersolar Ltd
Factory Three
Cock Lane
High Wycombe
UK

Italsolar
Via A. D'Andrea, 6
Nettuno 00048 (RM)
ITALY

KSB Pumpen
D-6710
Frankenthal (Pfalz)
GERMANY

Kyocera
Chiba-Sakura Plant
4-3 Ohsaku 1-Chome Sakura-Shi
Chiba-Pref 285
JAPAN

A Y Macdonald Manufacturing Co
4800 Chanvenelle Road
Dubuque
Iowa 52001
USA

Mono Pumps Ltd
Cromwell Trading Estate
Bredbury
UK

Mono Pumps Pty Ltd
338-348 Lower Dandenong Road
Mordialloc
Vic 3195
AUSTRALIA

Neste/NAPS
PO Box 96 Riuklokka
Brobekkveien 101
N-0516
Oslo 5
Norway

Photowatt International sa
131 Rt de l'Empereur
92500 Rueil-Malmaison
France

Photowatt Intnl S.A
65, Av du Mont Valerien
92500 Rueil-Malmaison
FRANCE

R & S Renewable Energy Systems
PO Box 45
5600 AA Eindhoven
THE NETHERLANDS
REIL
D-37 Madho Singh Road
Bani Park
Jaipur 302006

INDIA
Siemens Solar GmbH
Buchenallee 3
D-5060, Bergisch Gladbach
W GERMANY

Societe Nouvelle Chronar
3 Allee Edme Lheureux
Immeuble Vancouver
94340
FRNACE

Solar Energie Technik
Postfach 1180
D/6822 Altlusheim
GERMANY

Solarex Corporation
1335 Piccard Drive
Rockville
MD 20850
USA

Solarex Pty Ltd
78 Bibela Stree,
Villawood
PO Box 204
AUSTRALIA

Solar Jack International
c/o Energy Tech
13901 North 73rd Street
Scottsdale
Arizona 85260
USA

Southern Cross Int.
Box 454
Toowoomba
Queensland
Australia

Suntron
2/861 Doncaster Road
Victoria 3109
AUSTRALIA

Telefunken System Technik GmbH
Industriestrasse 23-33
D-2000 Wedel
Holstein
West Germany

Total Energie
24 Rue Joannes Masse
69009 Lyon
FRANCE

Zome Works Corporation
PO Box 25805
Albuquerque
NM 87125
USA

APPENDIX D : SPECIMEN SYSTEMS FOR GIVEN PUMPING SCENARIOS

Scenario 1 : Capacity 60 m³/day at 2 m head.

1. Chloride Solar
System: PF1/1
       Aquasol 100L pump from KSB.

Config: Floating
Rating: 318 Wp
Price : US $ 5390

2. IBC - (Using Kyocera)
System: 10 x 48 Wp modules from
        Kyocera (LA361J48)
        Rack, Wiring, Junction Box
        Aquasol 100L pump from KSB

Config: Floating
Rating: 500 Wp
Price : DM 13,400
        US $ 8320

3. A.Y. MacDonald & Co
System: 10 x MS5 modules from Arco
        Solar Inc., P.S. Permanent
        magnet d.c. motor, direct
        coupled. MacDonald S.P.
        centrifugal pump 830309DS

Config: Surface suction
Rating: 530 Wp
Price : On application

4. Kyocera (Sakura Plant)
System: SMB2-200-8/4
        D.c. motor.

Config: Unspecified
Rating: Unspecified
Price : 846,600 Yen
        US $ 5370

5. Mono Agricultural Division
System: Tracking, XMS07SP10T
        7 x 60 W modules,
        3/4 HP d.c. motor,
        1 x CP1600 pump from Mono.

Static: XMS10SP10
        10 x 60 W modules,
        1 HP d.c. motors,
        2 x CP1600 pumps from Mono.

Config: Surface suction
Rating: Tracking 420 Wp

Static 600 Wp
Price : A$ 11,000
        US $ 8430

6. Mono Pump Ltd.
System: MS12SP5 12 x 42 Wp modules
        2 x CP1600 Mono pumps,
        3/4 HP, permanent magnet
        motors.

Config: Surface suction
Rating: 504 Wp
Price : £ 5695
        US $ 9340

7. Suntron Pty.
System: XMS10
        10 x 60 Wp modules
        1 HP, 180 V d.c. motors
        800 W MPPT
        2 x CP1600 Mono pumps.

Config: Surface suction
Rating: 600 Wp
Price : $A 11,450
        US $ 8770

8. R&S
System: 8 x RSM40 modules,
        Aquasol 50M pump from KSB,
        brushless d.c., 1 stage

Config: Floating
Rating: 320 Wp
Price : NLG 11,200
        US $ 5880

Scenario 2 : Capacity 60 m³ at a head of 7 m.

1. Chloride Solar
System: PF 1/3
        KSB Aquasol 50M pump.
        Brushless D.C., 1 stage
        floating.

Config: Floating
Rating: 636 Wp
Price : US $ 8028

2. IBC (Kyocera)
System: 12 x 48 Wp Modules (Kyocera
        LA361J48) KSB Aquasol 50M
        pump. Rack, wiring and
        junction box.

Config: Floating
Rating: 600 Wp
Price : DM 15,250
        US $ 9020

3. A.Y. MacDonald & Co.
System: 14 x 55 Wp Modules (Arco M55)
        MacDonald S.P Centrifugal
        Pump 830309DS P.S. Permanent
        Magnet D.C. Motor, Direct
        Coupled.

Config: DC Submersible
Rating: 745 Wp
Price : On application.

4. Kyocera (Sakura)
System: SMD-750-18/6
        D.C. Motor

Config: Unspecified
Rating: Unspecified
Price : 1,419,140 Yen
        US $ 9010

5. Mono Agricultural Division
System: Tracking : XMS10SP10T
        10 x 60 W Modules,
        CP1600 Mono pump,
        1 h.p. d.c. motor.

Static: XMS12SP10, 12 x 60 W
        Modules, 2 x CP1600 Mono
        pumps, 1 h.p.d.c. motor.

Config: Surface suction
Rating: Tracking 600 Wp
        Static   720 Wp
Price : $A 12,660
        US $ 9700

6. Mono Pumps Ltd
System: MS16SP10
        16 x 42 Wp Modules
        2 x CP1600 Mono pumps,
        1 H.P D.C. permanent magnet
        motor.

Config: Surface suction
Rating: 672 Wp
Price : £ 6980
           US $ 11440

7. Suntron pty
System: XMS10
        10 x 60 Wp Modules
        2 x CP1600 Mono pump,
        1 H.P 180V D.C. motor,
        800W MPPT.

Config: Floating
Rating: 600 Wp
Price : $A 11,450
        US $ 8770

8. R&S
System: 16 x 40 Wp Modules , RSM40
        KSB Aquasol 50M pump.
        Brushless, floating, 1 stage
        D.C. pump.

Config: Floating
Rating: 640 Wp
Price : NLG 17,275
           US $ 9070

132

Scenario 3 : Capacity 20 m^3 at a head of 20 m.

1. Chloride Solar
System: PS 3/2
       Grundfos SP4-8
       A.C. Centrifugal

Config: AC Submersible
Rating: 742 Wp
Price : US $ 9377

2. IBC (kyocera)
System: 15 x 48 Wp Modules (Kyocera
       LA361J48)
       MacDonald 180810DP pump,
       Piping, rack, wiring,
       junction box,

Config: DC Submersible
Rating: 750 Wp
Price : DM 28,480
       US $ 16840

3. A.Y. MacDonald
System: 9 x 55 Wp modules (Arco M55)
       MacDonald 210008DM pump
       (Solar Sub)
       PAC SCI Brushless d.c.,
       w.p.m. coupled at max power.

Config: DC Submersible
Rating: 480 Wp
Price : On application

4. Kyocera (Sakura)
System: SMA-550G-18/18
       A.C. motor.

Config: Unspecified
Rating: Unspecified
Price : 1489640 Yen
       US $ 9460

5. Mono Agricultural Division
System: Tracking : XMS06BP20T
       6 x 60 W Modules,
       P301 Mono pump
       1/2 h.p. d.c. motor.

Static : XMS08BP20
8 x 60 W Modules,
P301 Mono pump,
3/4 h.p. d.c. motor.

Config: Surface-motor progressive
       cavity
Rating: Tracking 360 Wp
       Static   420 Wp
Price : A$ 10,000
       US $ 7660

6. Mono Pumps
System: MS10BP20
       10 x 42 Wp Modules,
       P301 Mono borehole pump,
       3/4 h.p. d.c. permanent
       magnet motor.

Config: Surface-motor progressive
       cavity
Rating: 420 Wp
Price : £5300
       US $ 8690

7. Suntron pty
System: XMS07
       7 x 60 Wp Modules,
       P301 Mono borehole pump,
       3/4 h/p/ 180 V d.c. motor,
       360 W MPPT.

Config: Surface-motor progressive
       cavity
Rating: 420 Wp
Price : $A 8930
       US $ 6841

8. R & S
System: 21 x 40 Wp Modules (RSM40),
       Grundfos SP4-8 pump,
       Submersible MS Centrifugal,
       A.C. motor,
       Inverter - Grundfos SA1500

Config: AC Submersible
Rating: 840 Wp
Price : NLG 25,835
       US $ 13,565

Scenario 4 : Capacity 40 m³ at a head of 40 m.

1. Chloride Solar
System: 2 x PS 2/4
       Grundfos SP2-18 pumps.

Config: AC Submersible
Rating: 2 x 1484 Wp
Price : US $ 31,931

2. IBC (Kyocera)
System: Kyocera LA361J48 Modules,
       Grundfos pump(s),
       Piping, racks, wiring.

Config: AC Submersible
Rating: Unspecified
Price : DM 58,000
       US $ 34,300 approx

3. A.Y. MacDonald
System: 30 x 45 Wp Modules (Arco
       M45),
       MacDonald 211008DM pump
       (Solar Sub),
       PAC SCi Brushless d.c.,
       w.p.m. coupled at max power.

Config: DC Submersible
Rating: 1410 Wp
Price : On application.

4. Kyocera (Sakura)
System: SMA2-1.1G-36/18
       A.C. motor.

Config: Unspecified
Rating: Unspecified
Price : 2,638,480 Yen
       US $ 16,760

5. Mono Agricultural Division
System: Tracking : XMS18BP40T
       18 x 60 Wp modules,
       P301 Mono borehole pump,
       1.5 h.p. d.c. motor.

Static : XMS28BP40
28 x 60 Wp modules,
P301 Mono borehole pump,
3 h.p. d.c motor.

Config: Surface-motor progressive
       cavity
Rating: Tracking 1080 Wp
       Static   1680 Wp

Price : Tracking $A 24,000
                 $ 18,390
       Static $A 28,000
                 $ 21,450

6. Mono Pumps
System: MS35BP40,
       35 x 42 Wp modules,
       P301 borehole pump,
       2 h.p. d.c. permanent magnet
       motor.

Config: Surface-motor progressive
       cavity
Rating: 1470 Wp
Price : £13,600
       US $ 22,300

7. Suntron pty
System: XMS28,
       28 x 60 Wp modules,
       P301 Mono borehole pump,
       3 h.p. d.c. motor,
       1800W MPPT.

Config: Surface-motor progressive
       cavity
Rating: 1680 Wp
Price : $A 26,650
       US $ 20,420

Figure E.1 : Breakdown of capital cost per Wp by configuration

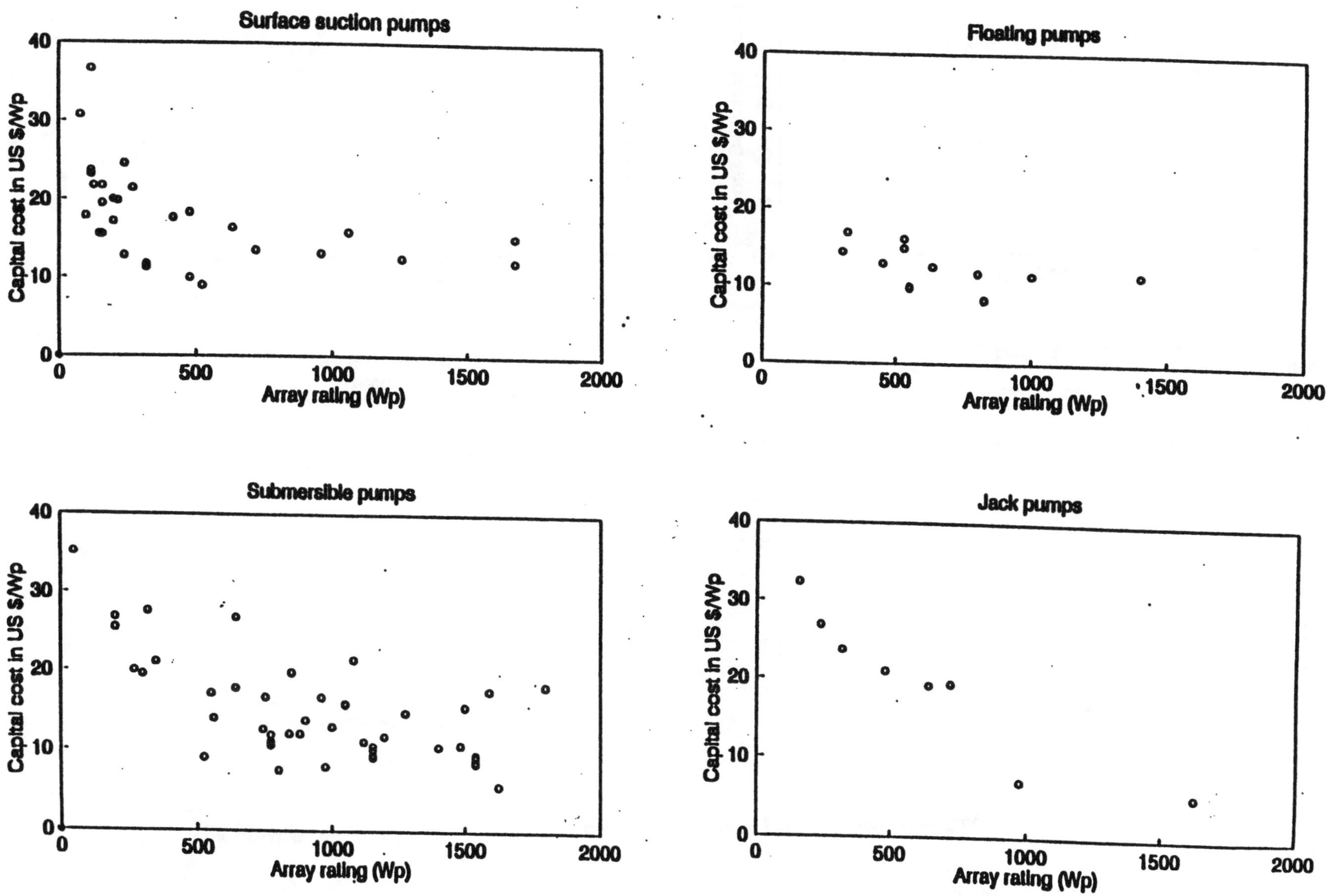

Figure E.2 : Breakdown of capital cost per m⁴ by configuration

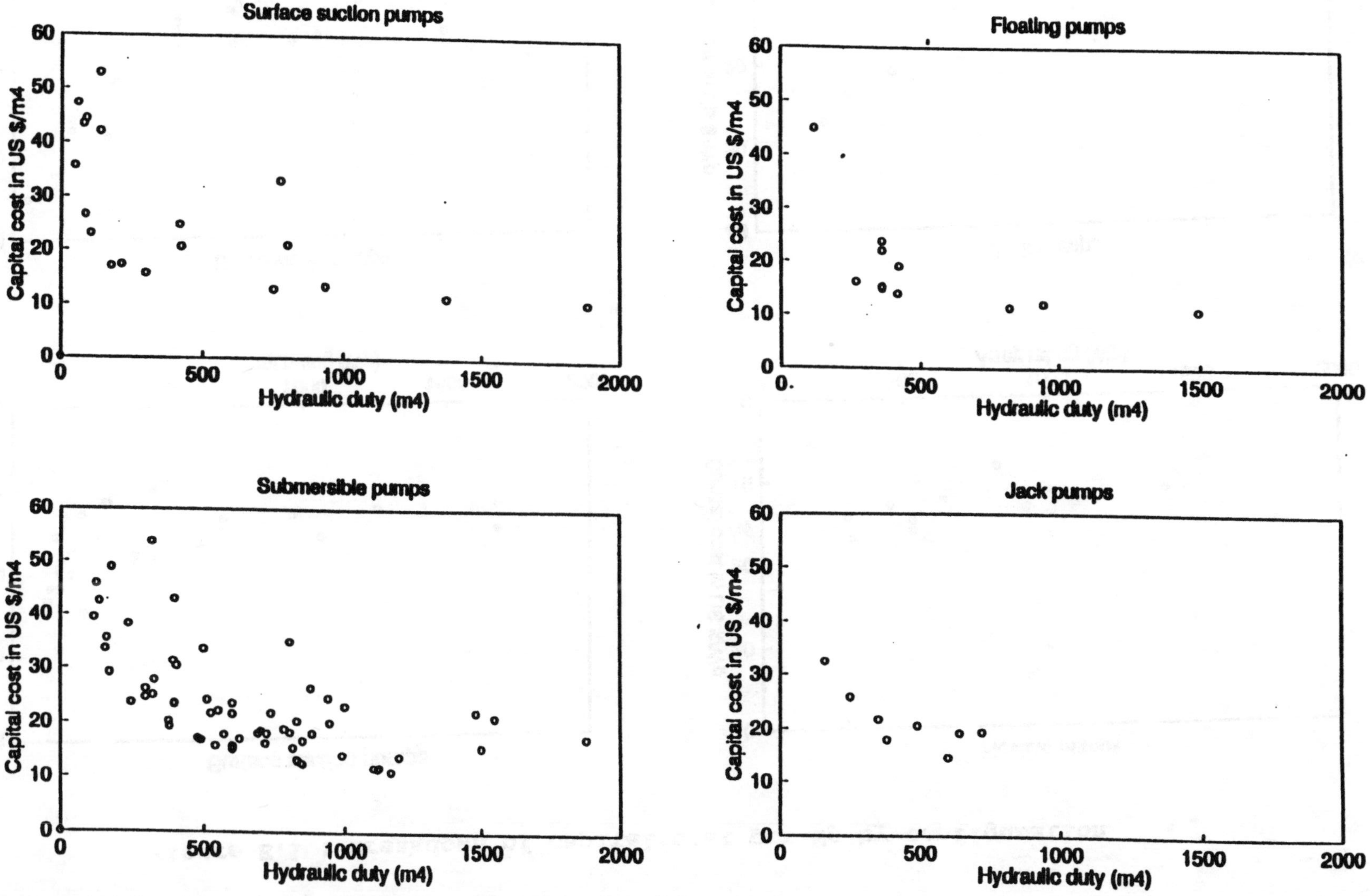

APPENDIX F : DISCOUNT FACTORS FOR LIFE-CYCLE COST ANALYSES

A description and examples of the use of these tables of discount factors in calculation Net Present Worth is given in section 8.1.2.

Table F.1 : Calculation of Discount Factor for a single future payment

Discount Rate (d)	Inflation Rate (i)	Factor Fs for given number of years				
		5	10	15	20	30
0.00	0.00	1.00	1.00	1.00	1.00	1.00
	0.05	1.28	1.63	2.08	2.65	4.32
	0.10	1.61	2.59	4.18	6.73	17.45
	0.15	2.01	4.05	8.14	16.37	66.21
	0.20	2.49	6.19	15.41	38.34	237.38
0.05	0.00	0.78	0.61	0.48	0.38	0.23
	0.05	1.00	1.00	1.00	1.00	1.00
	0.10	1.26	1.59	2.01	2.54	4.04
	0.15	1.58	2.48	3.91	6.17	15.32
	0.20	1.95	3.80	7.41	14.45	54.92
0.10	0.00	0.62	0.39	0.24	0.15	0.06
	0.05	0.79	0.63	0.50	0.39	0.25
	0.10	1.00	1.00	1.00	1.00	1.00
	0.15	1.25	1.56	1.95	2.43	3.79
	0.20	1.55	2.39	3.69	5.70	13.60
0.15	0.00	0.50	0.25	0.12	0.06	0.02
	0.05	0.63	0.40	0.26	0.16	0.07
	0.10	0.80	0.64	0.51	0.41	0.26
	0.15	1.00	1.00	1.00	1.00	1.00
	0.20	1.24	1.53	1.89	2.34	3.59
0.20	0.00	0.40	0.16	0.06	0.03	0.00
	0.05	0.51	0.26	0.13	0.07	0.02
	0.10	0.65	0.42	0.27	0.18	0.07
	0.15	0.81	0.65	0.53	0.43	0.28
	0.20	1.00	1.00	1.00	1.00	1.00

The actual formula used to calculate these factors, Fs, is

$$Fs = [(1 + i)/(1 + d)]^n$$

where i is the relative rate of inflation
      d is the discount rate
  and n is the number of years from the present

Table F.2 : Calculation of cumulative discount factor for annual payment

Discount Rate (d)	Inflation Rate (i)	Factor Fa for given number of years				
		5	10	15	20	30
0.00	0.00	5.00	10.00	15.00	20.00	30.00
	0.05	5.80	13.21	22.66	34.72	69.76
	0.10	6.72	17.53	34.95	63.00	180.94
	0.15	7.75	23.35	54.72	117.81	499.96
	0.20	8.93	31.15	86.44	224.03	1418.26
0.05	0.00	4.33	7.72	10.38	12.46	15.37
	0.05	5.00	10.00	15.00	20.00	30.00
	0.10	5.76	13.03	22.21	33.78	66.82
	0.15	6.62	17.06	33.51	59.44	164.68
	0.20	7.60	22.41	51.29	107.59	431.39
0.10	0.00	3.79	6.14	7.61	8.51	9.43
	0.05	4.36	7.81	10.55	12.72	15.80
	0.10	5.00	10.00	15.00	20.00	30.00
	0.15	5.72	12.87	21.80	32.95	64.27
	0.20	6.54	16.65	32.26	56.38	151.24
0.15	0.00	3.35	5.02	5.85	6.26	6.57
	0.05	3.84	6.27	7.82	8.80	9.81
	0.10	4.38	7.90	10.71	12.96	16.20
	0.15	5.00	10.00	15.00	20.00	30.00
	0.20	5.69	12.73	21.44	32.22	62.04
0.20	0.00	2.99	4.19	4.68	4.87	4.98
	0.05	3.41	5.16	6.06	6.52	6.87
	0.10	3.88	6.39	8.02	9.07	10.19
	0.15	4.41	7.97	10.85	13.18	16.58
	0.20	5.00	10.00	15.00	20.00	30.00

The actual formula used to calculate these cumulative discount factors, Fa, is :

$$Fa = \frac{\dfrac{(1 + i)}{(1 + d)} \dfrac{(1 + i)^n}{(1 + d)} - 1}{\dfrac{(1 + i)}{(1 + d)} - 1}$$

Where i is the relative inflation rate
      d is the discount rate
and n is the number of years for which the payment is made.

138

APPENDIX G: QUICK REFERENCE DATA FOR WIND, DIESEL AND HAND PUMPING

G.1 WINDPUMPING

Introduction to windpumps

This appendix is intended to give the reader the necessary information concerning windpumping to be able to perform an approximate life-cycle cost analysis. This is only for rough comparison with solar pumping, and should not be used in isolation as a justification for a installing a wind pump. Further reading on windpumping can be found in the 'Windpumping handbook' (see bibliography).

Windpumping is like solar pumping in that the resource level is an external factor, and the size of the collector (in this case the rotor) determines the extractable energy. Water storage is therefore necessary to provide supply in times of low mean winds. The wind resource is extremely site dependent on scales down to a few hundred metres, and it is therefore difficult to obtain reliable wind data from meteorological records. If possible, measurements should be taken at the exact site of the intended pump. Wind atlases are very general and can be misleading in this respect, but if these must be used, be sure that there is a suitably exposed location for the pump. The design month is that which requires the largest rotor to meet the demand. Windpumping can be effective at mean windspeeds of more than about 2.5 m/s, and is best suited to high head applications. This is because wind pumps are almost all reciprocating piston pumps. Windpump rotors are multi-bladed, giving a low rotation speed, but producing a high torque. They are thus less effective for high volumes of water at low heads. Windpumps need little maintenance, are robust and should have a turbine lifetime of 20 to 30 years. Windpumps can vary in rotor diameter between about 1m and 10m. Clearly, for the larger sizes installation is quite a major operation, and thus costs are correspondingly higher.

Sizing a windpump

The energy extractable is proportional to the area swept by the rotor. The windpump can therefore be sized by specifying its rotor diameter. The graph below (figure G.1) allows us to find this for different volume-head products and windspeeds: From the daily volume-head product on the horizontal axis, trace a vertical line up to the curve corresponding to the appropriate mean windspeed. Then trace a horizontal line towards the left to the vertical axis to read off A, the necessary rotor area. This defines the size of the windpump. Manufacturers usually specify their windpumps in terms of the rotor diameter, d. To find this from the rotor area, either use:

$$d = 2 \text{ x SquRoot } [A/Pi]$$

or use the left=hand half of the nomogram in figure G.1 by tracing across horizontally from the rotor area axis to the curve, and then tracing vertically downwards to read off the corresponding rotor diameter on the horizontal axis. The nomogram can of course be used in reverse to find the daily pumped volume given a certain rotor diameter and the mean windspeed.

Estimated windpumping costs

Capital cost	: 380 $/m2 of swept rotor area.
Pipework	: 5 $/m
Water storage	: 60 to 150 $/m3
Borehole drilling	: 60 to 200 $/m
Installation	: 20% of hardware capital cost
Rotor lifetime	: 20 to 30 years.
Pump lifetime	: 5 to 10 years
Pump cost (Repl)	: 300 to 800 $
Maintenance	: 2%/year of installed hardware cost
Operating costs	: None

Figure G.I: Nomogram for windpump sizing

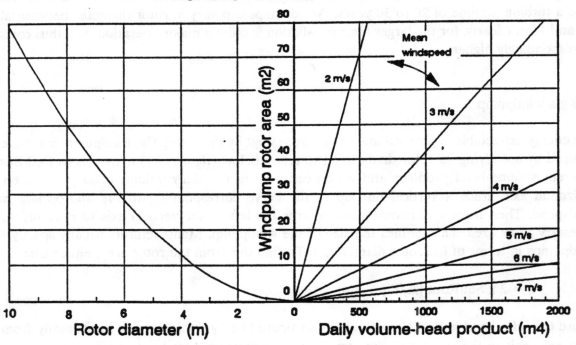

140

G.2 DIESEL PUMPING

Introduction to diesel pumps

This appendix is intended to give the reader enough information to be able to perform an approximate life-cycle cost analysis of diesel pumping. It is only included to give a rough comparison with solar pumping, and should not be used as justification for installing a diesel pump. Further reading on diesel pumping can be found in 'Water pumping devices', (see bibliography).

Diesel technology has the advantage that it is well known and understood all over the world, and so procurement, installation and maintenance are not such a problem. Diesel pumps will usually be oversized for most small applications, and so even the smallest, rated at about 2.5 kW (1.3 kW hydraulic power), will probably only need to be run for a short time each day. Short term water storage is not a necessity, as water can be pumped at any time on demand, but a tank is still advisable for village water supply applications. The design month is simply that with the highest water demand.

One of the main factors in deciding on the selection of a diesel pump is the cost, quality and availability of fuel. Supplies can often be spasmodic and unreliable in remote areas. Also, note that in an economic appraisal, the international market price of diesel fuel would be used. For a financial assessment the local price should be used, but this may vary widely depending on location.

An attendant is usually needed with a diesel pumping system, and maintenance needs to be frequent.

Sizing a diesel pump

As almost all diesel pumps will be oversized for small rural applications, we can select the smallest practical unit. Most pumps will have an overall efficiency between 6 and 9 %, and so the sizing essentially consists of calculating the daily fuel consumption. The graph in figure G.2 below enables us to calculate the number of hours of operation of a diesel pump as a function of volume head product. The second scale on the vertical axis gives a reading directly in litres of fuel used per day. This graph has been constructed for an engine using 1.5 litres of fuel per hour. Thus the number of litres of fuel per day can easily be found if the hydraulic duty in m4 is known. This can be turned into an annual fuel consumption figure.

Note: Diesel pumps are often rated in Horsepower, HP, where 1 HP is about 0.75 kW.

Diesel pumping costs

Capital Cost	: $600 to $1000 for a 2.5 kW pumpset
Installation	: 10% of diesel capital cost
Pipework	: 5 $/m
Storage (if necessary)	: 60 to 150 $/m³
Borehole drilling	: 60 to 200 $/m

Engine lifetime	: 10 years
Pump lifetime	: 10 years

Fuel (International price - non taxed) $0.30 per litre
Fuel (Local price) $0.50 to $2.00 per litre
Operation $1 per man day
Maintenance 200 $/year

Like operation and maintenance costs, fuel costs are an annual expenditure, and so its contribution to the total life-cycle cost must be calculated using the discount factors as described in section 8.1.2(b).

Many economists forecast that in future years the price of diesel fuel is likely to increase at around 3% above the general rate of inflation. Therefore you may wish to use i = 0.05 when finding the discount factor from table F.2 in appendix F.

Figure G.2: Nomogram for diesel pump sizing

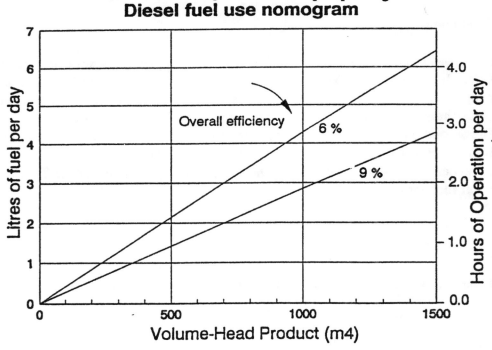

142

G.3 HANDPUMPING

Introduction to handpumps

Handpumps are the most widely used water lifting devices throughout the developing world. They are renowned for their reliability, simplicity and ease of repair using local technologies. However, their capacity is limited by the maximum rate of working sustainable by their human operators. Obviously, valuable man hours are also used in pumping water, and the cost of this labour must be attributed to the life-cycle cost of the pump. It is also worth noting that, in the case of handpumping from boreholes, when one pump is operating at maximum daily capacity, a further pump installation will often require another borehole to be drilled. This can be extremely costly, and in those circumstances other means of pumping are worth consideration. A shallow well may be able to support more than one pump. Except at very low heads and volume requirements, it is unlikely that handpumping will be suitable for irrigation purposes. Note that storage tanks are not normally used with handpumps, as water is pumped on demand.

Sizing the system

Handpumps tend to be of a standard size, which is, of course, related to the normal rate of working of its operator. Thus the system is effectively sized by determining how many persons are necessary to perform the required hydraulic duty. This can be calculated from the diagram in figure G.3. From the water lift (on the horizontal axis) and the daily volume requirement (on the vertical axis) the number of people required can be found. This assumes that a single person can produce 60 Watts of power for 4 hours a day, and that the handpump efficiency is 60%. Assuming the working day to be 8 hours this requires 1 extra pump for every two operators.

Handpump system costs:

Capital cost per pump:

15 m pump $700
25 m pump $1200
40 m pump $1900

Installation	: 12% of hardware capital cost
Lifetime of pump	: 5 years
Borehole Drilling	: 60 to 200 $/m
Maintenance	: 15%/year of installed hardware cost
Operation	: 1 to 2 $/man-day

Figure G.3: Nonogram for handpump sizing

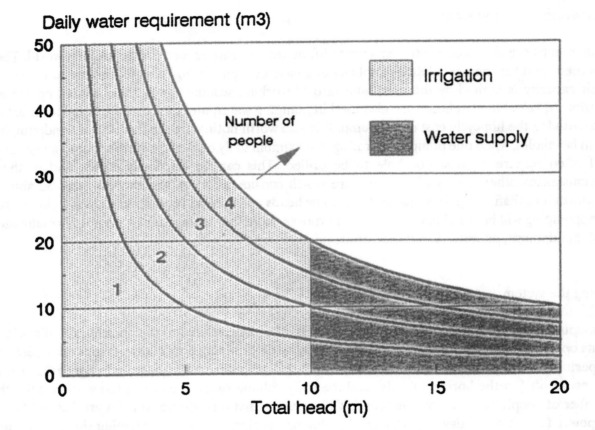

APPENDIX H: EXAMPLE OF A PV PUMP LIFE-CYCLE COSTING

This appendix gives an example of a life-cycle costing for a PV pump for village water supply. The situation we will use is that of the example sizing case worked through in section 6. Step one of the costing is therefore assumed complete. The analysis period will be taken as 20 years, the discount rate 10% per annum and the inflation rates of all items zero relative to general inflation. The basic scenario is as follows:

In April, the hottest driest month, the population using the pump is about 675, while in august, the coolest month, the number falls to 375. The details of per capita demand for each month are given in table 6.3. A new 6" borehole will be drilled to a depth of 55 m. From the analysis summarised in table 6.4 the necessary array size is 1000 Wp, to be used with an inverter and a submersible ac borehole pump.

Physical system layout: The array and tank are situated next to the pump and the tank is raised 2m off the ground. To provide 2 days storage in all but the highest demand month requires a tank of 30 m^3. To allow sufficient people to use the pump, a small distribution system is used with a main of length 15m branching into 4 stand pipes with taps. The total pipe length is 25m. The array, borehole and tank are enclosed in an area 20 x 15m by a simple fence to keep out animals, a total perimeter of legnth of 70m.

Step 2: Capital costs (guidance data from section 8.3.1)

The pumping system: Using table 4.2(a) for submersibles
we will assume a total system cost (array, inverter,
pump, support structure) of 12 $/Wp. Our system is
rated at 1000 Wp, hence system cost is: $ 12,000

Installation : For submersibles this is $250 + 0.6 $/Wp.
Thus for a 1000 Wp system the installation cost is
$250 + $600 = $ 850

Borehole drilling : We will take the drilling costs as
$100 per m. From the hydrological data example in
section 6.2 we defined our borehole depth as 55m. Hence
the borehole cost is: $ 5,500

Pipework rated at 5 $/m. We require 25m plus perhaps 5m
spare, so total plumbing cost is: $ 150

Storage tank : As this is relatively large, take a unit
cost near the bottom of the range, say 70 $/m3. For two
days storage in most months we require 30 m3. Hence

storage tank cost is:	$ 2,100
A simple fence should enclose the array, pump and tank. Estimated perimeter 70m. Assume a total cost of:	$ 100
Therefore total installed capital cost equals:	$ 20,700

Step 3: Recurrent costs (guidance data from sections 8.3.2 & 8.3.3)

Replacements costs: We will assume a pumpset cost of 2 $/Wp , so for our 1000 Wp system this is $2000. We must apply the discount factors for single future payments, as given in table F.1 in appendix F. We will use a discount rate of 10 %, giving a factor of 0.39 for 10 years. Thus:

Items	Cost	Replace After	Factor fs	P.Worth
motor/pump	$2000	10 years	0.39	$780
inverter	$1000	10 years	0.39	$390
Total lifetime replacement costs				$1,170

Annual costs: Maintenance is the only annual cost for solar pumping, which is estimated at about 1% of the installed system cost (excluding borehole, pipework and tank). This is therefore 1% of $12,850 which is about $130. The cumulative lifetime cost of this annual payment is found by multiplying by the factor fa from table F.2 in appendix F (using n = 20 years, i = 0, d = 0.1). This gives the factor fs as 8.51. Therefore the lifetime maintenance costs are 8.51 x $130 = $ 1,106

Step 4: Life-cycle costs

It now only remains to sum the capital cost and the future costs as follows:

Capital	$ 20,700
Replacements	$ 1,170
Maintenance	$ 1,106
Total life-cycle cost	$ 22,976

We can turn this figure into a yearly cost by dividing by the annualisation factor fa from appendix F table F.2, found in step 3 (which was 8.51 for 20 years).

Annualised Life-cycle cost (ALCC) = 2700 $/year

If the size of the community is constant (which in our example it is not) this can be worked out on a per capita or per household basis.

We should however calculate the unit delivered water cost: To do this find the total annual water use by summing the 'actual volume pumped' column in table 6.4 (206 m^3/day) and multiplying by 30 (the number of days per month) to give 6180 m^3/year. Then just divide the annualised life-cycle cost ($ 2700) by this figure to get the unit cost per cubic meter of water. Thus:

Annual water demand   = 6180 m^3
and so unit water cost   = 0.43 $/m^3$

It should be remembered that if a borehole already exists, unit water costs will be significantly less (in this case 0.32 $/m^3$).

APPENDIX I: THE ECONOMICS OF PV PUMPING

I.1 Comparative costs of solar pumping

In this section solar pumping will be compared to several other of the main pumping sources in the developing world: Diesel, hand and wind powered pumping. Because the specification for each pumping system depends so critically on the characteristics of the site, it is not possible to perform a life-cycle costing for a general case and definitively say that solar pumping will be the most cost effective solution given certain criteria. However, we can look at the comparative life-cycle costs over a broad but typical range of conditions, and make general implications about the types of scenario in which the different methods of pumping may be viable. It should be noted that this cannot take the place of a site-specific life-cycle costing, which is described in detail in section 8.

The principle of life-cycle analysis is to sum the total costs of a pumping system over its entire operational lifetime. This not only includes the initial capital and installation costs of pump, well and tank, but must include all future maintenance, fuel and component replacement costs. Future costs are multiplied by discount factors to account for expected changes in the value of money, and the opportunity cost of capital. In this way the cost of water from systems with a different balance of capital to operating costs, or different lifetimes can be compared against each other. The life-cycle cost can be expressed more meaningfully as a cost per annum, or as a cost per m^3 of water. The practicalities of the life-cycle costing method are described in detail in section 8, and discount factors are given in the tables in appendix F.

In this section we will make a general comparison of the economics of pumping for village water supply. A full life-cycle costing method will be used, using data typical of the developing world. Although they need not be explicitly stated here, the assumptions of costs and component lifetimes are based on actual figures wherever possible, and mostly correspond to the information given in section 8.3: 'Guidance on costs'. Irrigation pumping is not dealt with in this way, as the widely varying water demands of different crops and soil types makes a generalised approach impossible. Despite continuing reductions in real PV prices, PV irrigation pumping is still only suitable for a specialised range of criteria.

In the village water supply case analysed below, a typical set of conditions had been assumed, and the sensitivity of the unit water cost to varying village populations and water table depths examined.

Water lifting methods considered were: PV pumping; Windpumping; Diesel pumping with a directly connected mechanical pump; and Hand pumping. Each of the graphs in figures I.1(a,b,c & d) shows the way that unit water cost ($/m3) varies against village population for the four pumping methods under consideration. Village population sizes ranged from 100 to 2000 persons. The four graphs represent the situation for water depths 15, 25, 40 and 50 m respectively. Per capita water demand was assumed to be 40 litres per day, as targeted by the WHO. (It should be realised that the probable actual figure will be more like 20 litres per day at present where village water supply is by well-and-bucket or handpump).

In each case the system cost in $/Wp was calculated by using an exponential fit to the curve of manufacturers data in figures E.1 and E.2 (submersibles). The system size was calculated using the standard sizing algorithms used to produce the nomograms in section 6.5 (and appendix G for non-solar options).

Some of the other key parameters are listed below:

Daily per capita water demand	40 l
Days of storage	2
Design insolation	4.5 kWh/m^2
PV Subsystem efficiency	30%
Diesel efficiency	7%
Diesel fuel cost	0.30 $/l
Discount rate	10%/year

Several general points can be made from a brief inspection of the curves: PV pumping is always more economic than diesel pumping for low population numbers although this difference decreases as a higher pumping head becomes necessary. Conversely, diesel always becomes more economic at higher populations. The cross-over point where their unit water costs are equal corresponds to a village population of 400 at 15m head, falling to about 100 at 50m head. However, the difference in unit water costs is marginal over quite a large range, and for the 15m head case the two curves do not significantly diverge until a population of over 700.

Two wind power curves are shown, for mean winds of 3 and 4 m/s respectively. The 3 m/s curve is cheaper than both PV and diesel at very low populations, and tends towards the PV curve at higher populations. The 4 m/s curve is the lowest in all situations, except at depths greater than 40 and populations greater than 1000 persons, where it becomes marginally more expensive than diesel. Clearly wind energy can be very economic given high enough design month mean windspeeds.

Handpumps are generally available for heads of up to 40 m, and it is characteristic of handpumps that the unit water cost is constant with increasing population. At 15 m head handpumping is cheaper than PV for populations of less than about 500, but the cost increases rapidly with increasing water table depth. Handpumping is the most expensive option for all populations at water tables of 40 m and is regarded as unviable above this.

The analysis above has assumed a fixed economic price for diesel fuel of 0.3 $/litre. In reality the price can fluctuate greatly, from about 0.20 $/l to 1.50 $/l depending on taxes, subsidies, local availability and transportation costs. However, except at the upper end of this price range, the sensitivity to diesel fuel price is small, and does not affect the character of the results. Thus in a financial assessment using, say, 0.8 $/l, the population figure at which the PV and diesel cross may change quite considerably, even though the absolute value of the diesel unit water cost has changed very little. This is because the differences between PV and diesel are marginal over quite a large population range, and may be less than the inherent inaccuracy of a general analysis of this kind.

Figure I.1(a)   ## Water Pumping Cost Comparison
15m Water Table

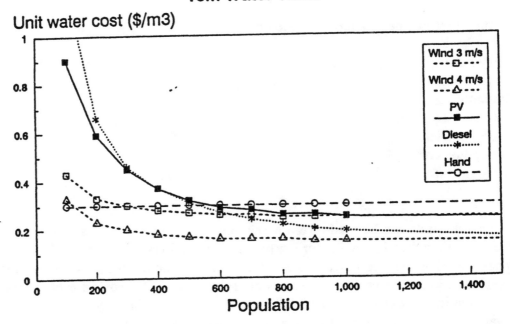

Figure I.1(b) ## Water Pumping Cost Comparison
25m Water Table

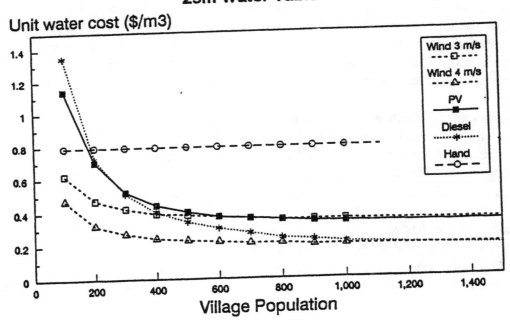

Figure I.1(c) **Water Pumping Cost Comparison**
40m Water Table

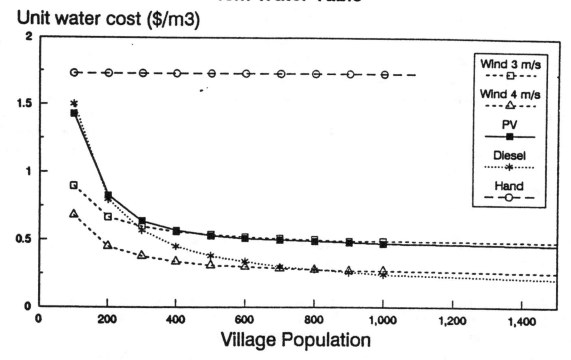

Figure I.1(d) **Water Pumping Cost Comparison**
50m Water Table

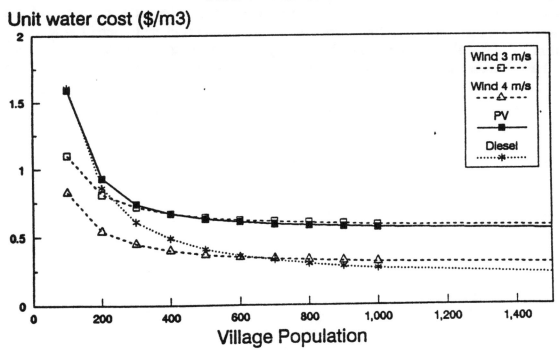

I.2 Projection

The two central factors that will affect the viability of PV pumping in the future are the world market price of diesel fuel, and the basic manufacturing costs of photovoltaic modules. Most of the other components in a pumping system are now well developed and standardised, and so large fluctuations in their cost are unlikely. Probably the most important of the two factors mentioned above will be the module cost, whose base value currently stands at 5 $/Wp. This is also the more predictable of the two, and a reasonable estimate of the relationship between module cost and future unit water cost can be derived. To illustrate this a sensitivity analysis was performed in which the module price was varied for a village of 500 people. Cases for water table depths of 15, 25, 40 and 50 m were examined. The unit water costs from this analysis are plotted in figure I.2. In the analysis it was assumed that the asymptotic value of the system cost for large systems (see figure E.1 for submersibles) is proportional to the base module price.

It can be seen that the unit water cost for a village of 500 persons drops by $0.08 for every 1 $/Wp reduction in module costs with a water table of 40 m, and by $0.04 for each $/Wp reduction with a water table of 15 m. With the prospect of new mass manufacturing methods reducing module costs to as low as 2 or 3 $/Wp during the next decade, PV is likely to become economic for a far wider range of applications. If this analysis holds true, then at 2 $/Wp, PV would be more economic than diesel over virtually the whole village water supply range. Timescales for this decrease in module costs are discussed in section 4.6. However, it is also likely that market factors will affect prices, and may prevent them from falling as low as technical factors might allow. It should also be borne in mind that previous projections have almost all been overly optimistic.

The question of diesel price fluctuations is more difficult to quantify, but as discussed in section I.1 above, quite large increases are necessary to significantly affect the character of the results. However, doubts concerning the security of oil supplies seem likely to persist at least for the first half of the decade, and this will almost certainly result in rising fuel prices. The only result of this can be a decrease in the viability of diesel in comparison to PV pumping.

Figure I.2: Sensitivity of unit water cost to module price

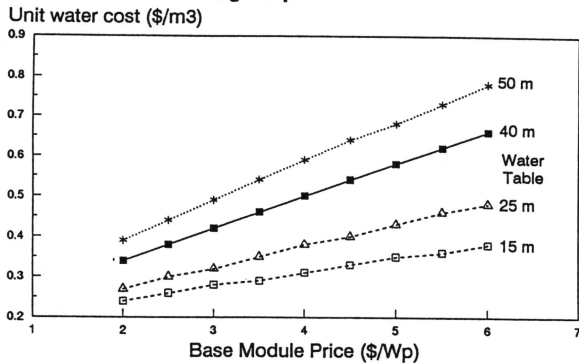

Distributors of World Bank Publications

ARGENTINA
Carlos Hirsch, SRL
Galeria Guemes
Florida 165, 4th Floor-Ofc. 453/465
1333 Buenos Aires

**AUSTRALIA, PAPUA NEW GUINEA,
FIJI, SOLOMON ISLANDS,
VANUATU, AND WESTERN SAMOA**
D.A. Books & Journals
648 Whitehorse Road
Mitcham 3132
Victoria

AUSTRIA
Gerold and Co.
Graben 31
A-1011 Wien

BANGLADESH
Micro Industries Development
   Assistance Society (MIDAS)
House 5, Road 16
Dhanmondi R/Area
Dhaka 1209

   Branch offices:
   156, Nur Ahmed Sarak
   Chittagong 4000

   76, K.D.A. Avenue
   Kulna 9100

BELGIUM
Jean De Lannoy
Av. du Roi 202
1060 Brussels

CANADA
Le Diffuseur
C.P. 85, 1501B rue Ampère
Boucherville, Québec
J4B 5E6

CHILE
Invertec IGT S.A.
Americo Vespucio Norte 1165
Santiago

CHINA
China Financial & Economic
   Publishing House
8, Da Fo Si Dong Jie
Beijing

COLOMBIA
Infoenlace Ltda.
Apartado Aereo 34270
Bogota D.E.

COTE D'IVOIRE
Centre d'Edition et de Diffusion
   Africaines (CEDA)
04 B.P. 541
Abidjan 04 Plateau

CYPRUS
Cyprus College Bookstore
6, Diogenes Street, Engomi
P.O. Box 2006
Nicosia

DENMARK
SamfundsLitteratur
Rosenoerns Allé 11
DK-1970 Frederiksberg C

DOMINICAN REPUBLIC
Editora Taller, C. por A.
Restauración e Isabel la Católica 309
Apartado de Correos 2190 Z-1
Santo Domingo

EGYPT, ARAB REPUBLIC OF
Al Ahram
Al Galaa Street
Cairo

The Middle East Observer
41, Sherif Street
Cairo

FINLAND
Akateeminen Kirjakauppa
P.O. Box 128
SF-00101 Helsinki 10

FRANCE
World Bank Publications
66, avenue d'Iéna
75116 Paris

GERMANY
UNO-Verlag
Poppelsdorfer Allee 55
D-5300 Bonn 1

HONG KONG, MACAO
Asia 2000 Ltd.
46-48 Wyndham Street
Winning Centre
2nd Floor
Central Hong Kong

INDIA
Allied Publishers Private Ltd.
751 Mount Road
Madras - 600 002

   Branch offices:
   15 J.N. Heredia Marg
   Ballard Estate
   Bombay - 400 038

   13/14 Asaf Ali Road
   New Delhi - 110 002

   17 Chittaranjan Avenue
   Calcutta - 700 072

   Jayadeva Hostel Building
   5th Main Road Gandhinagar
   Bangalore - 560 009

   3-5-1129 Kachiguda Cross Road
   Hyderabad - 500 027

   Prarthana Flats, 2nd Floor
   Near Thakore Baug, Navrangpura
   Ahmedabad - 380 009

   Patiala House
   16-A Ashok Marg
   Lucknow - 226 001

   Central Bazaar Road
   60 Bajaj Nagar
   Nagpur 440010

INDONESIA
Pt. Indira Limited
Jl. Sam Ratulangi 37
P.O. Box 181
Jakarta Pusat

ISRAEL
Yozmot Literature Ltd.
P.O. Box 56055
Tel Aviv 61560
Israel

ITALY
Licosa Commissionaria Sansoni SPA
Via Duca Di Calabria, 1/1
Casella Postale 552
50125 Firenze

JAPAN
Eastern Book Service
Hongo 3-Chome, Bunkyo-ku 113
Tokyo

KENYA
Africa Book Service (E.A.) Ltd.
Quaran House, Mfangano Street
P.O. Box 45245
Nairobi

KOREA, REPUBLIC OF
Pan Korea Book Corporation
P.O. Box 101, Kwangwhamun
Seoul

MALAYSIA
University of Malaya Cooperative
   Bookshop, Limited
P.O. Box 1127, Jalan Pantai Baru
59700 Kuala Lumpur

MEXICO
INFOTEC
Apartado Postal 22-860
14060 Tlalpan, Mexico D.F.

NETHERLANDS
De Lindeboom/InOr-Publikaties
P.O. Box 202
7480 AE Haaksbergen

NEW ZEALAND
EBSCO NZ Ltd.
Private Mail Bag 99914
New Market
Auckland

NIGERIA
University Press Limited
Three Crowns Building Jericho
Private Mail Bag 5095
Ibadan

NORWAY
Narvesen Information Center
Book Department
P.O. Box 6125 Etterstad
N-0602 Oslo 6

PAKISTAN
Mirza Book Agency
65, Shahrah-e-Quaid-e-Azam
P.O. Box No. 729
Lahore 54000

PERU
Editorial Desarrollo SA
Apartado 3824
Lima 1

PHILIPPINES
International Book Center
Fifth Floor, Filipinas Life Building
Ayala Avenue, Makati
Metro Manila

POLAND
International Publishing Service
Ul. Piekna 31/37
00-677 Warzawa

For subscription orders:
IPS Journals
Ul. Okrezna 3
02-916 Warszawa

PORTUGAL
Livraria Portugal
Rua Do Carmo 70-74
1200 Lisbon

SAUDI ARABIA, QATAR
Jarir Book Store
P.O. Box 3196
Riyadh 11471

**SINGAPORE, TAIWAN,
MYANMAR,BRUNEI**
Information Publications
   Private, Ltd.
Golden Wheel Building
41, Kallang Pudding, #04-03
Singapore 1334

SOUTH AFRICA, BOTSWANA
For single titles:
Oxford University Press
   Southern Africa
P.O. Box 1141
Cape Town 8000

For subscription orders:
International Subscription Service
P.O. Box 41095
Craighall
Johannesburg 2024

SPAIN
Mundi-Prensa Libros, S.A.
Castello 37
28001 Madrid

Librería Internacional AEDOS
Consell de Cent, 391
08009 Barcelona

SRI LANKA AND THE MALDIVES
Lake House Bookshop
P.O. Box 244
100, Sir Chittampalam A.
   Gardiner Mawatha
Colombo 2

SWEDEN
For single titles:
Fritzes Fackboksforetaget
Regeringsgatan 12, Box 16356
S-103 27 Stockholm

For subscription orders:
Wennergren-Williams AB
Box 30004
S-104 25 Stockholm

SWITZERLAND
For single titles:
Librairie Payot
1, rue de Bourg
CH 1002 Lausanne

For subscription orders:
Librairie Payot
Service des Abonnements
Case postale 3312
CH 1002 Lausanne

TANZANIA
Oxford University Press
P.O. Box 5299
Maktaba Road
Dar es Salaam

THAILAND
Central Department Store
306 Silom Road
Bangkok

**TRINIDAD & TOBAGO, ANTIGUA
BARBUDA, BARBADOS,
DOMINICA, GRENADA, GUYANA,
JAMAICA, MONTSERRAT, ST.
KITTS & NEVIS, ST. LUCIA,
ST. VINCENT & GRENADINES**
Systematics Studies Unit
#9 Watts Street
Curepe
Trinidad, West Indies

TURKEY
Infotel
Narlabahçe Sok. No. 15
Cagaloglu
Istanbul

UNITED KINGDOM
Microinfo Ltd.
P.O. Box 3
Alton, Hampshire GU34 2PG
England

VENEZUELA
Libreria del Este
Aptdo. 60.337
Caracas 1060-A